U.S. Options for Energy Independence

U.S. Options for Energy Independence

Edited by
Nake M. Kamrany
University of Southern California

LexingtonBooks
D.C. Heath and Company
Lexington, Massachusetts
Toronto

Library of Congress Cataloging in Publication Data

Main entry under title:

U.S. options for energy independence.

Includes index.
1. Energy policy—United States. I. Kamrany, Nake M., 1934–
II. Title: US options for energy independence.
HD9502.U52U18   333.79'0973   81–48394
ISBN 0–669–05361–9   AACR2

Published simultaneously in Canada

Printed in the United States of America

International Standard Book Number: 0–669–05361–9

Library of Congress Catalog Card Number: 81–48394

Contents

Foreword

On the occasion of its Centennial, the University of Southern California is pleased to present this book based upon the proceedings of a conference entitled "Solutions to the Energy Problem" held on campus July 11, 1980.

Rising energy costs in recent years have permeated almost every aspect of our lives. We face ever-increasing bills for transportation, illumination, and heating. At the same time, the cost of energy is affecting virtually everything we buy. This is a result of business enterprises passing on to the consumer their increased production costs caused by high-priced energy.

This upward pricing spiral represents a dramatic reversal in mankind's history. As a feature of our progress, there has been a continuous decline in the cost of energy from the animal power of the ancient world to the power of the gasoline combustion engine in the last century. Possibly history may record this meteoric rise in the price of energy as the most significant event in the last quarter of the twentieth century.

It was out of concern about the far-reaching significance of the energy price increase that the motivation for this conference arose. The Newport Foundation, a group of public-spirited citizens from Newport Beach, California, headed by Dr. Delmar Bunn, a physician, proposed that the University of Southern California sponsor a conference to discuss the economics of alternative energy sources. The objective was to bring together representatives of the Foundation with a group of economists from industry, government, and academe. All of these individuals have devoted immeasureable amounts of their time and talents addressing the energy issues of our nation. The format of this book was developed in consultation with Professor Nake Kamrany of the USC Economics Department, who has arranged many conferences for USC and other groups on economic matters.

In the book, industry is represented by papers from business economists and technologists of Union Oil Company and Getty Oil. Government is represented by economists from the Library of Congress, the Department of Energy and the Rand Corporation, a private organization involved in research for the Government. And the viewpoints of academicians come from the University of Southern California, as well as the California Institute of Technology, The Massachusetts Institute of Technology, Stanford University, and the University of California at Los Angeles.

The fact that the participants did not reach a consensus on specific recommendations for a course of action merely reflects our nation's current travail as it attempts to weigh the advantages and disadvantages of oil, coal, biomass, natural gas, solar and atomic energy while it agonizes over dependence on oil imports as opposed to the higher cost of producing domestic energy.

Economists maintain that a solution to a problem is generally not a choice between either this or that, but rather less of this and somewhat more of that. The conference, however, does appear to have reached the conclusion that the United States must reduce its dependence on foreign energy sources. The same conclusion drawn earlier by Dr. Bunn and the Foundation had prompted them to approach USC regarding the conference.

A series of decisions lies before us to determine the extent of our dependence on energy imports and domestic production, and it is my hope that the deliberations of the experts at the conference may make some contribution in shaping our energy policy of the future.

James H. Zumberge
President
University of Southern California

Editor's Note:

This volume is designed in response to a need for synthesizing and disseminating authoritative views about the U.S. national energy issues to a wide audience—students, policy makers, and the general public. An attempt is made to find congruence between one of the significant issues of energy—stockpiling, deregulation, import quotas, OPEC, alternative supplies, and macroeconomic effects—and one or more scholars, industry practitioners, public policy makers, and the citizenry group.

The energy issue is a typical economic problem. It is not well understood and the formation of policies to deal with it is fraught with difficulty. This is partly due to the fact that the energy issue is exceedingly complex and multi-dimensional; it has a dynamic character that influences and is influenced by the economy, and it transcends national boundaries. For example, the price of energy is influenced by the demand of the domestic economy and the decisions of OPEC; it contributes to domestic inflation and is influenced by the general price level by de-facto indexing, and it has become a major element of the balance of merchandise trade with supernational implications.

The specific focus of the volume, however, is to expound upon the various dimensions of alternative solutions and specify policy prescriptions related to the national energy issue.

Since the inception of the energy crisis in 1973/74, a number of major studies have been made, many bills have been proposed and passed, numerous conferences have been held, and various administrations have proposed and implemented their energy programs. As a result, an enormous amount of information has been generated on the social, technological, and economic aspects of energy. Although we are not sure that an optimal solution is in sight, nevertheless, we feel that some serious deliberation on the specific aspects of a probable solution(s) is warranted at this time.

It is our expectation, at least, that this volume will deepen and enrich the opportunities for serious intellectual debate on the U.S. energy problem and will contribute to a better understanding and insight into concepts, principles, problems, and prospects of the energy issues. Needless to say, this volume, like many others, will not provide a final word on its subject nor an unassailable point of view.

The optimistic tone of "solutions" to the energy crisis is predicated on the conviction that the American ingenuity and dedication will solve the energy problem, given the necessary resources and time, if the American public is convinced that a real problem exists and is willing to confront it. While many years of research have been carried out and millions of words have been written on various aspects of the United States energy issue, there is a dearth of lucid publications on the policy problem and its possible solutions that would elucidate the issue clearly to the average citizenry as well as the policy makers. This volume is designed to close this gap by inviting noted researchers and accomplished industry practitioners to address the "solution" issue without technical jargon. To our good fortune, I believe, the contributors ably addressed themselves specifically to problem identification and elucidation of alternative solutions. The chapters are brief, succinct, lucid, and yet comprehensive. A number of the contributors have spent decades on energy research and have distinguished themselves with scholarly contributions and technological innovations in the field of energy.

This book consists of the proceedings of a conference which was held at the University of Southern California in July of 1980, entitled, "Solutions to the Energy Problem: Alternative Approaches to the United States Energy Crisis." The idea of the conference and the publication of the proceedings originated with my good friend and collaborator, Dr. Delmar Bunn, the founder and president of the Newport Foundation for the Study of Major Economic Issues. Indeed, I have found Dr. Bunn to be a stimulator of ideas, a first rate organizer, and his relentless drive for tackling major economic issues and his genuine concern for seeking solutions for major economic issues have transformed the ideas of the Foundation into a national forum. I was further enlightened by the active participation and deliberations of the members of the Foundation, my students and colleagues at USC in the deliberation of the energy issues. I am most grateful to the contributors of this volume who so effectively responded to my invitation and were equally generous to my demands for details, deadlines, revisions, and reviews.

Gratefully acknowledged are the financial contributions of the following public-minded organizations whose grants speeded up the publication of this volume:

      Newport Foundation for the Study of Major Economic Issues
      American Medical Associates
      Petrolane Corp.
      Signal Oil Company
      Union Oil Company
      Pacific Mutual Insurance

I also wish to express my deep gratitude for the support provided by the Department of Economics at the University of Southern California. Enthusiastic staff support was provided by Ida Abby, Janet Meritt, Margaret Mondoza, Linda Anderson, Vivian Smith, Sally Barnes, Mabel Solares; Kathy Haines assisted with the bibliography and related research; Lily Kamrany designed the art work. Copy editing, final proofing, and index were the responsibility of Janet Taylor. As always, I enjoyed the moral support and intellectual guidance of my colleague and collaborator, Professor Aurelius Morgner of the Department of Economics at U.S.C.

At home Sajia, Wali, Lily, and Michael were generous with their time, support, love, and affection and to whom I am deeply indebted.

Nake M. Kamrany
Pacific Palisades, Calif.

Observations on the Reagan Energy Plan

INTRODUCTION

The title of this book and a majority of the proposals contained herein reveal a clear bias for U.S. independence in energy. There are two underlying reasons: One, energy is unique as compared to thousands upon thousands of other goods and services that are being transacted among millions of consumers and producers daily. It should be noted that we are not necessarily advocating U.S. independence in gold, silver, bread, paper, . . . etc. Energy, however, is unique because the engine of production depends upon it. U.S. imports of oil constitute a huge sum ($80-$100 billion) which in a significant way affects domestic inflation and the U.S. balance of trade and payments. Most of the existing supply of energy is exhaustible posing a potential problem in the long run. While short term solutions via the market mechanism or otherwise may postpone the long term problem, they will not solve it, because the long term solution would depend upon technological developments in the energy field, and these technological developments require substantial investment in research and development which now embodies a high risk and a longer, than what an investor considers reasonable, payback period.

The supply of imports has been and will remain subject to disruptions for political reasons. The oil price spiral has been disruptive and inflationary. Prices are largely determined by the OPEC cartel. The demand for certain types of energy such as oil is *relatively* inelastic, especially in the short run, although effective user conservation is being observed in response to the oil prices of 1980 and 1981.

Secondly, in view of the uniqueness of energy, the U.S. is in an advantageous position to turn a problem or crisis into a major opportunity. It follows that the time horizon (short, medium, and long term) of the energy market is of critical importance. The non-renewable fossil fuel based energy of the present will have to be substituted by renewable science-based energy of the future. The *linkage* between "the present" and "the future" energy market requires the development of technologies which are subject to *risk* and *a long waiting period* for investment payoff. Generally, investors' behavior and preference under the market system is characterized by myopic optimization, i.e., profit maximization with a short investment payback rule. The market system is not a new invention. There are now some 200 years of experience with the market system. It performs perfectly well for short term optimization if left alone. Even then, the market system is subject to fluctuations and discontinuity. In the oil market, the supply side is controlled by a cartel that would manipulate prices for its economic and political advantage. Moreover, the development of these technologies require on the average 20 + year of lead time.

It follows that the private investor would have to have the *incentives* to take such a risk for such a long lead time. How can the proper environment be created for the private investor? With this brief background, the Reagan Administration energy policy is examined briefly.

II. THE REAGAN ADMINISTRATION ENERGY POLICY

The Reagan Administration energy policy was submitted to Congress in July of 1981, and was entitled, "Securing America's Energy Future: The National Energy Policy Plan, A Report to the Congress Required by Title VII of the Department of Energy Organization Act-Public Law 95-91".

The essence of the plan is the expression of the Administration's general principles and guidelines concerning energy. Many of the specific measures are awaiting further review. Nevertheless, the general guidelines manifest themselves in the Administration's fundamental philosophy of adherence to a free market system with heavy reliance upon the private sector responses—both the investors and the consumers—to market incentives. Moreover, the formulation of the energy guidelines adhere to the Administration's overall Economic Recovery Program which is now popularly known as "supply-side economics".

The general tenure of the "guidelines" deserves high grades since it forcefully articulates the energy problem and confronts it with flexibility and a dynamic approach. The reformulation represents a major departure from the previous administrations in a number of areas as briefly discussed below.

A. Domestic Energy Production

It is recognized that the Federal Government controls one third of the land area of the United States which contains an estimated 40% of uranium, 35% of coal, 85% of tar sands, 80% of oil shale, 40% of natural gas, 80% of oil, and 50% of the nation's geothermal resources. While the specifics are not spelled out yet, "the Federal role in national energy product is to bring these resources into the energy marketplace, while simultaneously protecting the environment." Richard J. Stegemeier makes a similar plea, "Reopen Federal lands for oil and gas", and provides detailed quantitative information in support of his claim in Chapter 4 (pp. 29-41).

B. Reliance Upon the Market

On January 28, 1981, President Reagan issued an Executive order decontrolling the price of petroleum and petroleum products immediately. This action was indeed in the right direction. Nearly all of the major studies on this issue had advocated decontrolling the domestic price of petroleum (see Chapters 1-4, 6-7, 9, 14, and 15 in this book). President Reagan speeded up the process.

However, the Administration has not taken a similar measure with regard to decontrolling prices of natural gas. Moreover, the Administration is working toward a Strategic Petroleum Reserve (SPR) of 750 million barrels by 1989. In the case of SPR, the Administration has argued that "commercial enterprises have no economic reason to achieve stockpile levels that are optimal from the national perspective" (p. 19). Although the Administration's position is in consonance with the recommendations of Walter S. Bear (see Chapter 12), it certainly is a departure from relying upon the private sector. If the SPR provides a legitimate exception of the Administration's overall philosophy, then one could argue that there may be other exceptions as well especially in regard to the creation of new energy technologies.

C. Regulatory Reform

In addition to decontrolling the price of petroleum and abolishing the Department of Energy, the President's Task Force on Regulatory Relief is focusing a comprehensive review on existing and proposed regulations in the petroleum and other energy areas. The Administration's action on reducing reporting burdens throughout the industry has already saved 700,000 man-hours annually. Dave Button of Getty Oil has pointed out (see Chapter 14 below) that the "petroleum industry has had to bear a relatively greater injection of inefficient government regulation and taxation over a ten-year period than has any other major industry in the history of this country." In conformance with Executive Order 12291, decisions about particular regulations will be based on four principles: (1) limiting Federal involvement; (2) increasing agency accountability for regulatory actions; (3) putting greater emphasis on cost-effectiveness in setting regulatory requirements; and (4) generally reducing regulatory burdens on the private sector. (p. 23).

No doubt the Administration's regulatory reform will have a net benefit to the nation and on this issue the Administration deserves high marks.

D. Oil Imports

The Administration's position on oil imports is vague. On the one hand, it expresses a desire for "efficient displacement of imported oil . . .", on the other it argues that market forces should not be distorted through "indiscriminate subsidies for alternatives that cost more than imported oil now . . .".

It appears that the Administration's view of the "cost" of imported oil is distorted if it compares the nominal cost of imported oil (e.g., $34/b) to the cost of alternatives that are generated domestically. To the nominal cost of imported oil, one must add the additional cost of stockpiling, since SPR is designed to guard against import disruption, i.e., 750,000,000 barrels x $34/b at 20% interest rate. Moreover, the cost of import disruption has other dimensions. In addition to its explicit cost ($34/b + $x/b for SPR), there are many implicit costs such as domestic economic dislocations, massive unemployment, economic depression, and the possibility of drawing U.S. military response. These costs could be actually much greater than the explicit cost of imports. Moreover, a reduction in oil imports through an optional tariff will reduce the cost per barrel, a pecuniary externality that is being ignored by the Reagan Plan. This could amount to around $10/b—a substantial sum! All other importers, Western Europe, Japan, and the LDC, will benefit from the United States action regarding imports.

E. Final Recommendation

Realistically speaking, at this time the U.S. should adopt a two-tier pricing policy for petroleum: a domestic price that is set much higher than the imported price in order to encourage conservation and the development of alternatives, and an import price much below $34 per barrel ($24/b), in order to reduce the burden of oil imports upon the balance of trade payments. Since the U.S. is a major purchaser, world prices will eventually conform to the U.S. established price. This can be accomplished by putting a ceiling on the total dollars for oil imports as recommended by Delmar Bunn in Chapter 5 below. It should also be noted that in the absence of an OPEC cartel and under free market conditions the price of a barrel of oil should be equal to its marginal cost, i.e., below $30/b. The difference between the domestic price and the imported price should be used for the development of alternative sources of energy or could be used as a national energy dividend which is proposed by Prof. Michael Intriligator in Chapter 8 below.

In addition, we must create *incentives* for the private sector to engage in R ± D investment for producing alternative sources of energy that will eventually bring about energy independence for the U.S. One optimal way to create the necessary incentives is to guarantee prices and markets for new energy. Details of such an option are contained in Professor Lester Thurow's proposal in Chapter 3 below.

It is not clear how a strategy of energy independence for the U.S. will cause damage to other free world economies or in any way reduce the U.S. posture as a reliable trade partner.

A strategy for energy independence will reduce insecurity for the U.S. and other importers due to import disruption. More importantly, it will enable the U.S. to retain its position as the leader of the free world. Just in the same way that the government helped create the computer which has helped the private sector enormously, it could do the same for energy independence.

With respect to synthetic fuels and other nonconventional sources of energy, the issue is not with respect to the commercialization of the technologies of alternative fuels but with respect to the creation of the technologies. The Administration guideline states, "The use of renewables should continue a healthy growth as the rising cost of conventional fuels and the new tax incentives stimulate demand" (p. 12). The assumption of rising cost of conventional fuels is dubious. As Lester Thurow has pointed out, OPEC may set its prices to preempt the development of alternatives. Why must the U.S. let OPEC dictate the timing, quality and quantity of the development of alternative sources of energy?

The Administration plan states, that the Federal Government recognizes a direct responsibility to demonstrate the scientific and engineering feasibility of nuclear fusion! The same argument could be made for solar, synfuel, and other renewables. However, the budgets have been drastically cut for synfuel and the renewables, while a shift is observed in favor of nuclear energy. Is solar as an alternative dying because of these cuts? Are the development of alternative fuels being abandoned?

In summary, while the Reagan plan is to be applauded for its regulatory reforms and the introduction of efficiency into the governmental measures concerning energy, its absence of a strategy of energy independence and not reducing oil imports for the long run, requires some hard rethinking.

Nake M. Kamrany
Santa Monica, Calif.

1

The Vicious Circle of the United States Energy Problem

Nake M. Kamrany

I. INTRODUCTION

Many views have been expressed concerning the origin and evolution of the United States energy problem. As the old rubic goes, there are as many diagnoses of the energy problem as there are economists—each pointing a finger at the villian who created it. In this paper, I have argued that the roots of the problem lie in the creation of huge aggregate demand for energy in the United States and substitution of oil for other sources of energy, leading to dependence upon imports. Moreover, a series of events—price controls, OPEC cartels, supply manipulation, the Middle East wars, Soviet competition, risk and uncertainty of supply of oil, and myopic optimization behavior—have led to a major problem whose cumulative effect is identified below in Figure 1-1. This cumulative effect is the vicious circle of the energy problem.

II. AGGREGATE DEMAND FOR ENERGY

The aggregate demand for energy evolved from the 1860's as an outgrowth of a series of short-run, rational behaviors depicting typical myopic optimization responses to a number of economic opportunities on the part of those who were the consumers of energy (households and businesses) and those who supplied energy to the market (oil companies). To be sure, our decision makers over the years have ignored the longer view of the energy issue. Instead, the energy problem has resulted from a sequential and interacting chain of events over the last century involving many factors, including resource discoveries, substitution of other sources of energy with oil, and a lack of courage by many administra-

1

tions in adopting a long-term, consistent and realistic policy. While it is difficult to ascertain a clear cut causality, all of the above factors in varying degrees contributed to a substantial rise in the energy intensity and per capita consumption of energy in the United States which resulted in an enormous aggregate demand for energy. United States per capita energy consumption is three times greater than Japan and twice as much as Western Europe, as shown in Table 1-1.

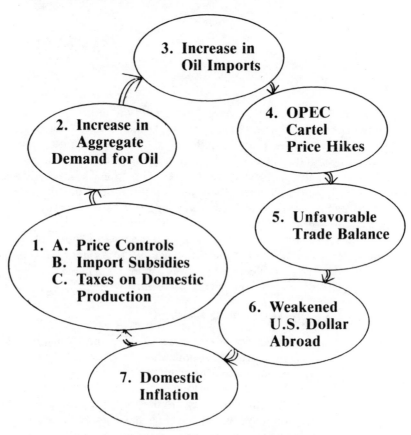

Fig. 1-1. The Vicious Circle of The U.S. Energy Problems.

Thus, after nearly a century of this panacea, we are caught in a vicious circle of the energy problem which is rooted in the growth of the aggregate demand for energy with large oil imports. Clearly, the energy problem was in the making long before the 1973-74 Arab oil embargo. The embargo was the symptom and not the cause of the problem.

There are those who believe that if the market forces were left alone, the free interaction of demand and supply would have brought about an equilibrium condition of the energy market. Energy prices determined by the market would have cleared any deficits. Higher prices for energy would have encouraged domestic production of oil and technologies for alternative sources of energy would have been developed. An equilibrium condition, by definition, means that buyers and sellers are satisfied at the market prices and quantities offered, and there would be no tendency or desire for change.

The above classical argument, despite its simplicity and static nature, would have been palatable if the market mechanism had not been tinkered with and if the energy market had lent itself to the free forces of a competitive market on the demand and supply sides of the equation. However, the energy market, which was dominated by oil, represented a market structure of bi-lateral monopoly. On the supply side, the oil producing countries initially had a weak, unorganized and non-collusive oligopoly which eventually turned into a strong, organized, and collusive oligopoly (cartel). On the demand side, the oil companies which were initially strong, organized and collusive at the international level (for example the Seven Sisters) became unorganized, non-collusive, and weak. Eventually, the oil companies, having lost their bargaining position vis-á-vis the oil producing-countries, have taken the role of royalty or tax collectors for OPEC. Moreover, the behavior of oil supply prices since 1970 has been nonconventional in the sense that instead of small incremental changes, major price jumps have created shocks and greater supply risk and uncertainty of imports.

Table 1-1. Total[1] Energy Consumption Per Capita

(million metric tons of coal equivalent)

(Figures in parenthesis are annual percent growth rates)

	USA[1]	Canada[2]	Japan	Fed. Republic of Germany	France	United Kingdom
1978	11.474 (−0.9)	9.9297 (−0.5)	3.8247 (0.5)	6.0139 (4.0)	4.3677 (8.8)	5.2029 (1.9)
77	11.574 (0.7)	9.9791 (0.2)	3.8056 (0.6)	5.7829 (−2.2)	4.0138 (−7.5)	5.1035 (2.5)
76	11.497 (5.7)	9.9612 (2.1)	3.7819 (4.4)	5.9123 (9.3)	4.3403 (10.7)	4.9800 (0.1)
75	10.874 (−4.3)	9.7575 (−0.1)	3.6214 (−5.4)	5.4085 (−5.0)	3.9223 (−9.4)	4.9801 (−4.0)
74	11.357 (−3.7)	9.7650 (1.4)	3.8298 (−2.3)	5.6937 (−2.3)	4.3283 (−2.8)	5.1895 (−4.3)
73	11.789 (1.8)	9.6364 (1.8)	3.9218 (11.2)	5.8282 (4.8)	4.4546 (4.8)	5.4235 (1.0)
72	11.577 (3.9)	9.4669 (5.2)	3.5269 (2.1)	5.5615 (2.1)	4.2514 (4.6)	5.3690 (−1.8)
71	11.146 (2.6)	8.9951 (3.3)	3.4440 (4.0)	5.4460 (2.2)	4.0631 (2.6)	5.4694 (1.7)
70	10.855 (1.0)	8.7076 (3.8)	3.3115 (15.8)	5.3263 (6.2)	3.9595 (8.3)	5.3754 (4.3)
69	10.745	8.3889	2.8604	5.0151	3.6575	5.1552

1) Solid fuels, liquid fuels, natural gas, hydro, nuclear, electricity
2) Data include imports of natural gasoline not shown elsewhere

Sources:

Consumption:	"World Energy Supplies 1950-1974", U.N.
	"World Energy Supplies 1973-1978", U.N.
Population:	"Demographic Yearbook 1977", U.N.
	"Yearbook of Labor Statistics 1979"
	International Labor Office
	"World Statistics in Brief 1979", U.N.

Thus the source of U.S. oil imports is now characterized by the existence of risk and uncertainty. Risk means the probability function of foreseen events such as the substantial OPEC price hikes, while uncertainty refers to the probability function of unforeseen events such as the Imam Khomeini revolution. This foreseeability concept depends upon the time horizon and the quality of information. Prior to 1973-74, the probability of an Arab oil embargo was not foreseen, nor was the war between Iraq and Iran foreseen a decade ago. It is disconcerting to note that U.S. oil import rates increased substantially beginning in 1970. Eighty-four percent of all U.S. oil imports are from OPEC.

The existence of risk and uncertainty in the supply of oil poses a major national security and economic problem because of the tremendous aggregate demand for oil imports. While this demand has grown steadily over time, the supply of oil has fluctuated over time due to a number of exogenous factors or factors that are outside the control of the domestic economy.

III. **OIL SUBSTITUTION**

From the mid-1800's until 1973-74, the relationship of the relative prices of alternative sources of energy evolved in such a way that gas and oil, which were most expensive in the mid-1800's, became less and less expensive over time. By the mid-1900's gas and oil became the least expensive in terms of relative prices (see Figures 1-2 through 1-6). As a result wood and coal were replaced by oil. By 1980, oil constituted 50% of the source of energy consumption in the United States, as illustrated in Figures 1-6 and 1-7 and as shown in Table 1-2. Moreover, the cheap price and environmental consideration of oil pre-empted the development of alternative sources of energy, such as coal, hydro, nuclear, and non-conventional sources such as solar, geothermal, wind and others. In Table 1-3 the distribution of the consumption of energy by sector is shown.

Our high dependency upon oil as a major source of energy has three significant drawbacks. 1) Exhaustible Resource—oil is an exhaustible resource and sooner or later (say, 30 to 50 years from now) we would have to find alternative sources of energy. While the issue of depletion of oil poses a long term problem which may be ameliorated by the development of alternative energy sources, oil has presented to us an immediate and short-term problem. Since half of the United States' oil consumption is being imported, the producing countries have manipulated the availability of the supply of oil for both political non-price reasons and economic reasons, i.e., the producing countries wish to maximize the present value of a long-term stream of revenues from oil, extracting ine highest possible price the traffic can bear.

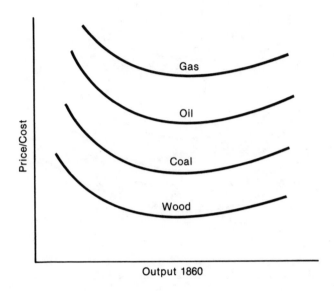

Figure 1-2. Relative Production Cost: Sources of Energy, 1860

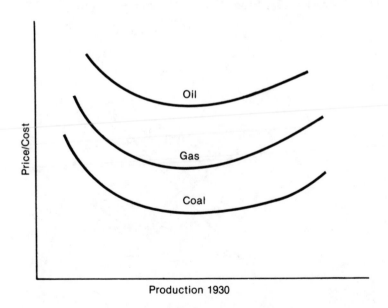

Figure 1-3. Relative Production Cost: Sources of Energy, 1930

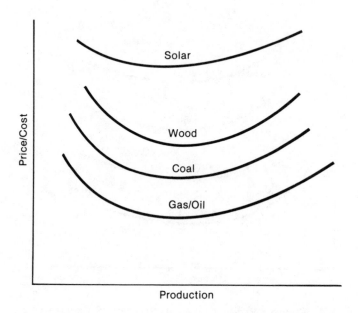

Figure 1-4. Relative Production Cost: Sources of Energy, 1980

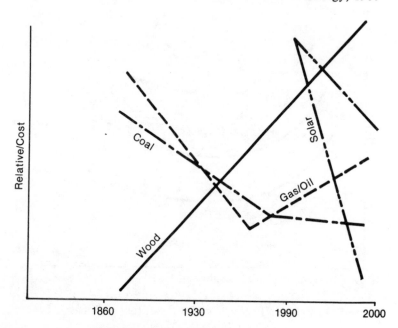

Figure 1-5. Trends in Relative Costs of Alternative Sources of Energy

Figure 1-6. U.S. Percentage of World Consumption of Energy

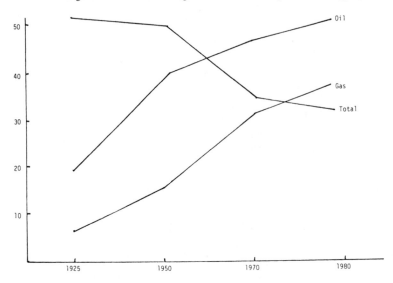

Figure 1-7. Percentage of Energy Consumption in the United States by Source of Energy

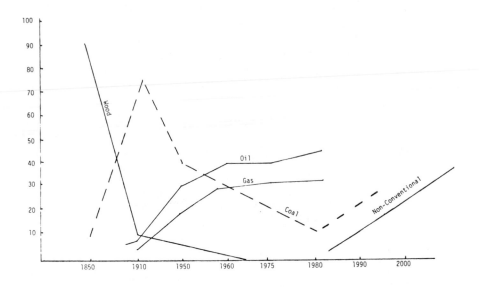

Table 1-2. **Sources of Energy Consumption, 1980**

Petroleum	50%
Gas	25%
Coal	18%
Nuclear	4%
Hydro	3%
Non-Conventional	0%

Source: Department of Energy, 1980

Table 1-3. **Consumption of Energy in the U.S. (Quadrillion Btu)**

	Residential & Commercial	Industrial	Transportation	Total
1979	29.503 (38%)	28.746 (37%)	19.766 (25%)	78.022
78	28.941 (37%)	28.581 (37%)	20.625 (26%)	78.154
77	28.268 (37%)	28.362 (37%)	19.753 (26%)	76.390
76	27.933 (37%)	27.495 (37%)	19.074 (26%)	74.509
75	26.743 (38%)	25.763 (36%)	18.195 (26%)	70.707
74	26.800 (37%)	27.895 (38%)	18.058 (25%)	72.759
73	27.559 (38%)	28.518 (38%)	18.526 (25%)	74.609
72				71.643
71				68.698
70				67.143
69				64.979

Sources:

"Monthly Energy Review", Department of Energy;
"The Energy Index 1974"

2) Unreliable Source of Supply — the 1973-74 Arab oil embargo and the 1979 Iranian oil disruption are illustrative of the political conflicts and employment of power as a means of obtaining political concessions and shifting the economic rent for the production and consumption of oil from the consuming to the producing countries. Moreover, the emergence of nationalism and its expression in the so-called North-South dialogue has become recriminatory, hostile, and unpredictable toward the West. Regional and border conflicts such as the Iraq-Iran war of 1980-81 point to supply disruptions in the future and the unreliability of the supply of oil at any price. According to a Library of Congress study, if the vital Strait of Hormuz were to close during the Iraq-Iran war, the price of oil would soar to $100/barrel at the 1980 equivalent prices due to shortages. This factor is indeed beyond the control of any nation state, especially since the Middle East region has become so volatile. The Soviet invasion of Afghanistan also points to the geopolitical superpower competition and the possibility of a Soviet role in the supply disruption of oil for the West.

3) Import Bill — the formation of the OPEC cartel as an organized and collusive oligopoly has made it possible for the oil-producing countries to limit the supply and *up the price* of oil. This has significant implications upon the domestic economy and the trade sector. For instance, in 1980, the import bill for oil is estimated at 100 billion dollars. That is a significant figure even in a $2.5 trillion GNP. It should be noted that the dollar value of oil imports as a percentage of the GNP was less than half of one percent in 1972, it is around 4% now, and if the OPEC price trends are moderate, it will be 6% by 1985. It means that the cost of oil imports will be about the same as or greater than the total value added of the U.S. agricultural output, or the total value added of the construction industry.

Moreover, since OPEC knows that the price of oil will increase further in the future, it attempts to make less oil available, thus saving more for the future markets at higher prices.

The traditional view of economists about cartels' behavior is that, in due time, they disappear and generally only cause a short term inconvenience because members of a cartel, sooner or later, cheat on each other by lowering prices in order to improve their relative share of the market. Such behavior leads to retaliation by other members of the cartel who then lower their prices also. This chain of events eventually leads to a price war and break up of the cartel. However, the creation of the oil cartel was made possible largely because of the growing aggregate demand which the United States largely created for oil. The U.S. imports 37% of the total world exports. As long as the market demand grows, the existence of the cartel is assured since its members can sell more oil if they wanted to at higher instead of lower prices. This market situation has prevented retaliation (lowering of prices by other members of the cartel) against the initial cheater.

The United States policy with respect to oil has indeed contributed to the sustenance of the OPEC cartel. Eighty-four percent of U.S. oil imports are from OPEC and by following a policy of cheap domestic oil, i.e., providing ample supply of oil at prices as low as possible, the demand for oil imports soared. It is true that this "cheap oil" policy has had its benefits. It contributed in a significant way to the economic growth and well-being of the nation. The upsurge in technological progress, a seven-fold increase in real per capita GNP over the last century, and the development of huge markets were partly due to the substitution of oil for other sources of energy. However, the growth of this sizable aggregate demand for oil has locked the economy, at least in the short run, into dependence upon unreliable sources of oil. What was not anticipated — or, if it was anticipated, it was ignored — has turned out to be a big monster.

It is somewhat ironic that the signals of the 1973/74 oil embargo did not alert the government into seeking effective conservation measures and increased production of domestic oil, the development of alternatives to oil, and less dependence upon oil imports. The time lag for countermeasures were ignored. For instance, it takes 5-10 years to adopt effective conservation measures such as the development of public transportation systems or the adaptation of less energy-intensive production. The development of new technologies for alternative sources of energy takes about 20 years. We have lost precious lead time. In fact, the record of the 1973-80 period reveals that the government continued with a policy of cheap energy, continued with a high dependency upon oil as a major source of energy, continued to encourage oil imports and discouraged domestic production of oil, and did not follow a vigorous program of developing new sources of energy. From 1973 to 1980, U.S. domestic production of oil declined by 7%. During the same period, net oil imports increased by 35%. However, the 1980 data reveals an 18.4% drop in oil consumption from 1979, largely due to higher prices. Also, domestic oil exploration in 1980 was at a historic high in response to higher prices and expectation of even higher prices in the future.

IV. PRICE CONTROL AND IMPORT SUBSIDIES —
MARKET INTERVENTION

To illustrate the domestic market intervention, let us examine the price controls-entitlements program for crude oil that was instituted in 1974, and our long-standing system of price controls on natural gas wellhead prices. The crude oil price controls-entitlements program has basically worked to tax domestic production of oil, and the proceeds have been used to subsidize imports. Under the program, the producer must purchase an "entitlement" at a cost (e.g., $2/barrel) in order to refine domestic crude oil while refiners who import crude oil at the world price received an entitlement per barrel which amounted to an import subsidy.

This policy increased the demand for oil from OPEC. In a realistic sense, the policy contributed to the strength of OPEC. OPEC, in turn, increased its prices further — more than a tenfold increase has occurred since 1973. Moreover, by maintaining oil prices domestically below world market prices, the policy contributed to increased consumption far above what the market would have allowed, thereby increasing the necessity of further imports. Clearly, as imports grew, greater strains were placed on the tax and subsidy programs. Although the entitlement program is being phased out, it has had a major impact upon the creation of the vicious circle of energy crises in the United States.

The regulation of natural gas field markets by the Federal Power Commission (FPC) is another illustration of the government's policy to keep the price of energy low. The FPC has for many years held the wellhead price of gas far below the world market level. This policy contributed to a demand growth of 5.3% per year, while domestic production began to decrease in 1972 since incentives for producers to explore for and extract natural gas were taken away. Shortages began to show and, according to the FPC, the 1976-77 shortage was 23% of "firm" requirements. Persistent shortages of natural gas have contributed to about 2 million barrels per day of imports. Although the 1978 Act will rectify these defects, the consequences of these policies are still being felt.

Why have such policies been followed? Perhaps there are many answers to this question. The overriding consideration, however, has been the *distributional objectives.* in other words, the government has attempted to prevent a redistribution of income and wealth from energy consumers to energy producers. This policy has failed. The record shows that there has been a substantial shift of the economic rent to the energy producers and to OPEC, as exemplified by the huge profits of the oil companies and the surpluses that OPEC has amassed.

The nation would have been better off if the government had not used energy as a means for its distributional objective, since there are more direct and more efficient means to reach that objective. The consequences of tinkering with the market forces, i.e., setting prices below market prices, can be illustrated even at an elementary level of economics. As shown in Figure 1-8 below, the free interaction of demand (D) and supply (S) for oil barring intervention would have an equilibrium price (Pe) and an equilibrium quantity (Qe). Whenever prices are controlled and set below market price, such as the United States price control of domestic oil shown as Pc, it leads to larger quantities, Qc, of oil being consumed than the equilibrium quantity of Qe. Thus, the price control, as illustrated, has contributed to an excessive increase in the consumption of oil from Qe to Qc. And an oil consumption at Qc has made it possible for OPEC to raise prices up to Po.

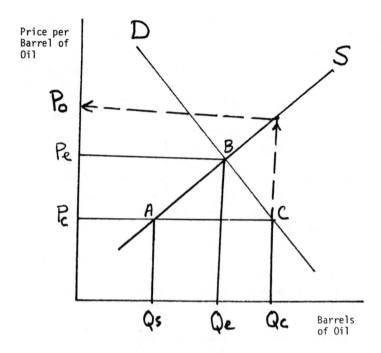

Figure 1-8. The equilibrium market price, Pe, is determined where the demand, D, for oil and the supply of oil, S, are equal, i.e., where they intersect and Qe barrels of oil is consumed. However, at a control price of Pc, Qc barrels of oil is consumed, which is substantially greater than the equilibrium quantity Qe. At the control prices, Pc, the suppliers are willing to sell only Qs. Under no market intervention, there would be a supply deficit of ABC. At Qc consumption, a price of Po would have to be offered.

V. Policy Rationale

A number of misconceptions about the nature of the energy market added to the confusion regarding energy policy. These included the following fallacies: the fallacy of low price elasticity of demand for energy; the fallacy of a high growth elasticity of energy upon the GNP and economic dislocation, misconceptions about rigidities of technological development of alternative sources, including nonconventional sources, externalities, voluntary and involuntary conservation.

An important point to note is that the above policies of price controls and import subsidies have already created an enormous aggregate demand in an ex-post-facto sense which is now difficult to satiate or scale down. It is simply difficult to modify one's lifestyle, shake off the old habits of Sunday driving or not give a car for a 16th birthday gift. The automobile has become a major instrument in our entertainment, life style, status, and social habits. Moreover, the designs of our homes and our cities are made upon the assumption of the availability of cheap oil. How are we going to undo all of that which evolved over decades? The design of the city of Los Angeles and Southern California is really based upon the assumption of cheap oil. The 1980 Chrysler loss of over $1.2 billion and losses by all other U.S. auto manufacturers of $4.2 billion are the direct result of the "cheap energy" policy and the failure of these auto manufacturers to anticipate the "energy crunch" and adjust their auto size and fuel efficiency accordingly.

VI. THE INFLATION EFFECT

Oil has become a major item in the international sector of the U.S. economy since the import bill is the largest single item in the ledger of the international merchandise trade. More importantly, however, there is a significant relationship between the prices of internationally traded goods (imports and exports) and domestic prices (traded and non-traded goods). Moreover, the price of tradeable goods such as oil indirectly affect the prices of non-tradeable goods. In view of the fact that the price of oil has been increased more than 15 times the 1973 prices, a very rapid rate of increase to say the least, and that oil figures rather importantly in the United States foreign sector, it has precipitated significant weight upon the price of tradeable and non-tradeable goods and, thus, upon the rate of domestic inflation in the United States.

Although oil imports are a small percentage of the GNP, one would assume that a big price increase in oil would have a small impact on the U.S. economy. This would be an inappropriate way to measure the impact of oil prices on the U.S. economy because approximately 25% of the goods included in the consumer price index and 50% of the goods included in the wholesale price index are determined in world markets rather than in U.S. markets. In other words, the impact of a rise in oil prices on the U.S. economy is much greater than would be suggested by looking at oil import shares of the GNP.

In summary, it has been argued that over the last century oil was substituted for other sources of energy in the U.S. consumption of energy (50% of the total). A conscious policy of "cheap energy" created a heavy dependency upon foreign sources of oil and created an enormous aggregate demand for imported oil (per capita U.S. energy consumption is nearly twice that of West Germany). This strong aggregate demand for oil from imports contributed to the formation, creation, and continuance of the OPEC cartel. In turn, the OPEC cartel employed its monopoly power both to exact political concessions and to shift the economic rent from the production and distribution of oil from the consuming countries (the United States consumes 44% of total free world oil production and 37% of its exports) to the producing countries. The OPEC price hike has contributed $100 billion to the import bill in 1981 which has weakened the U.S. dollar abroad, caused an unfavorable balance of merchandise trade, and contributed in a significant way to domestic inflation. And finally, domestic inflation feeds into aggregate demand for oil. This entire process is reiteration and was illustrated in Figure 1.

Clearly, we are caught in this vicious circle of the energy problem at a time when political upheavals in the Middle East (the Iraq-Iran war of 1980, or any unilateral decisions by the major oil suppliers, or a Soviet invasion of Iran, Saudi Arabia, or a number of these countries) could cause major oil supply disruptions. What can we do? We must break the vicious circle.

A response will be most appropriate if we draw a clear distinction between the immediate and short run problem versus the long run solution. In the long run (10-20 years and beyond if we provide the incentives now), we could confidently rely on technologies for the development of alternative sources of energy as illustrated in Figure 5 provided that sufficient collaborative private/public research-and-development investment is made in technologies of alternative energies. We must be cognizant of the time factor here — on the average, it takes around 20 years from the inception of an idea to its marketability if the technological and economic feasibility criteria are met. In the short run, we need to reduce our aggregate demand of energy through involuntary conservation measures and substantially reduce our dependence on foreign oil (increase domestic production of oil) and seek means of minimizing the impact of oil supply disruption through appropriate international arrangements.

A fundamental solution for the short and long run would call for matching or rather synchronizing our aggregate demand for energy to the growth of the economy and our carrying capacity of domestic energy resources. The implication of this approach may impose major readjustment and tightening of the belt in the short run. For the long run, one can never tell what technological developments, inventions, and innovations in the energy field hold for us. The time requirements and complexities of creating alternative sources of energy must not be underestimated. If the past is a reliable guide, then we have no reason to give up optimism. For instance, with the possibility of solar energy becoming a major source of energy supply, we may have discovered a source free from geopolitical consideration, an inexhaustible, clean source for short run use and a source that is cheap and clean in the long run. That possibility will end the vicious circle of the energy problem and make the U.S. a major exporter of energy in the future.

On January 28, 1981, President Ronald W. Reagan ordered an immediate decontrol of oil prices. This is indeed a necessary step in the right direction in terms of breaking the vicious circle of the energy problem. It must be supplemented with policies to control the dollar outlays for oil imports coupled with policies to augment the development of alternative sources of energy.

Selected Bibliography

Allen, Edward L. *Energy and Economic Growth in the United States.* Perspectives in Energy Series. Cambridge, Mass., and London: MIT Press, 1979.

Anderson, R. G. *Energy Conservation and Factor Substitution in U.S. Manufacturing.* Ph.D. Dissertation, Department of Economics, M.I.T.: Sept., 1979.

The American Assembly, Columbia University. *Energy Conservation and Public Policy.* Englewood Cliffs, N.J.: Prentice Hall, 1979.

Banks, F. E., *The Political Economy of Oil.* Lexington, Mass.: D.C. Heath and Co., 1980.

Bemis, Virginia. *Energy Guide: A Directory of Information Sources.* Assisted by Denton E. Morrison et al New York and London: Garland. 1977.

Berndt, E. R., et al., Dynamic Models of the Industrial Demand for Energy, Report EA-580, Palo Alto, Calif.: Electric Power Research Institute, 1977.

Bohi, Douglas and Russell, Milton. *Limiting Oil Imports: An Economic History and Analysis.* Baltimore, Md.: Johns Hopkins University Press, 1978.

Bohi, Douglas and Russell, Milton. *U.S. Energy Policy: Alternatives for Security.* Resources For the Future, 1975.

Brannon, G.M. *Studies in Energy Tax Policy.* Cambridge, Mass.: Ballinger, 1975.

Department of Energy. *Model Documentation, 1979 and Short Term Energy Outlook, Report AR/IA/80, 1980. Report,* Energy Information Administration, 1980.

Doran, Charles F. *Myth, Oil and Politics: Introduction to the Political Economy of Petroleum.* New York, N.Y.: Macmillan, 1977.

Dox, Samuel A. *Energy: A Critical Decision for the U.S. Economy.* Grand Rapids, Mich.: Energy Education Publishers, 1977.

Ericson, Edward W. and Waverman, Leonard. *The Energy Question: An International Failure of Policy.* Buffalo, N.Y.: University of Toronto Press, 1974.

Ezzati, Ali. *World Energy Markets and OPEC Stability.* Lexington, Mass.: D.C. Heath and Co., 1978.

Ford Foundation and Resources for the Future. *Energy: The Next Twenty Years.* Cambridge, Mass.: Ballinger, 1979.

Gilliland, Martha W. ed. *Energy Analysis: A New Public Policy.* Boulder, Colo.: Westview Press for the American Association for the Advancement of Science, Washington, D.C., 1977.

Griffin, J.M. *Energy Conservation in the OECD: 1980-2000.* Cambridge, Mass.: Ballinger 1979.

Hoch, I. *Energy Use in the United States by State and Regions,* Washington, D.C.: Resources for the Future, 1978.

Hudson, E. A. and Jorgenson, D.W. *"The Economic Impact of Policies to Reduce U.S. Energy Growth," Discussion Paper 644,* Cambridge, Mass.: Harvard Institute of Economic Research, August, 1978.

Knowles, Ruth S. *The American Oil Famine: How It Happened and When It Will End.* New York, N.Y.: Coward, McCan and Geoghegan, 1975.

Laird, Melvin R., et al. *U.S. Energy Policy: Which Direction?*. Moderated by John C. Daly. EI Forum, no. 8 AEI Roundtable held on June 27, 1977. Washington, D.C.: American Enterprises for Public Policy Research, 1977. Pp. 45.

Landsberg, Hans H. et al. Energy: *The Next Twenty Years: Report by a Study Group Sponsored by the Ford Foundation and Administered by Resources for the Future.* Cambridge, Mass.: Ballinger, 1979.

Lawrence, R. *New Dimensions to Energy Policy.* Lexington, Mass.: D.C. Heath and Co., 1980.

Mangone, Gerald J., ed. *Energy Policies of the World, Volume III.* New York, N.Y. and Oxford: Elsevier, 1979.

Mead, Walter J. and Deacon, Robert T. *Proposed Windfall Profits Tax on Crude Oil: Some Major Errors in Estimation, The Journal of Energy and Development,* Autumn, 1979.

Medvin, Norman. *The Energy Cartel: Who Runs the American Oil Industry?.* New York, N.Y.

Nordhaus, William D. *The Efficient Use of Energy Resources.* Cowles Foundation for Research in Economics at Yale University, New Haven, Conn. and London; Yale University Press, 1979.

Nordhaus, William D., ed. *International Studies of the Demand for Energy: Selected Papers Presented at a Conference in the International Institute for Applied Systems Analysis, Austria.* Assisted by Ramy Goldstein. Contributions to Economic Analysis Vol. 120.: New York and Oxford: distributed in the U.S. by American Elsevier.

Phillips, Owen. *The Last Chance Energy Book.* Baltimore, Md. and London: Johns Hopkins University Press, 1979.

Pindyck, Robert S. *The Structure of World Energy Demand.* Cambridge: M.I.T. Press, 1979.

Pindyck, Robert S. (ed.), *Advances in the Economics of Energy and Resources,* Conn.: JAI Press, 1979.

Russel, Joe W., Jr. *Economic Disincentives for Energy Conservation.* Environmental Law Institute State and Local Energy Conservation Project, Cambridge, Mass.: Harper & Row, Ballinger, 1979.

Schaffer, E.H. *The Oil Import Program of the United States.* New York: Praeger, 1968.

Schelling, Thomas C. *Thinking Trough the Energy Problem.* New York, N.Y.: Committee for Economic Development, 1979. Pp. xi, 63.

Schurr, Sam H. *Energy in America's Future.* Baltimore, Md.: Johns Hopkins University Press, 1979.

Stobaugh, Robert and Yergin, Daniel. *Energy Future.* New York, N.Y.: Random House, 1979.

United California Bank. *The Energy Crisis: Its Implications for the U.S. . . . Its Impact on California.* Los Angeles, Cal.: Research and Planning Division, United California Bank, 1979.

Vernon, Raymond. *The Oil Crisis.* New York, N.Y.: Norton, 1976.

Watkins, G. Campbell and Walker, Michael. *Oil in the Seventies: Essays on Energy Policy.* Vancouver: Fraser Institute, 1977.

Wilson, C.L. (Project Director). *Energy: Global Prospects, 1985-2000.* Report of the Workshop on Alternative Energy Strategies. New York, N.Y.: McGraw-Hill, 1977.

2

Looking for Villains in the Energy Crisis

Roger G. Noll

Americans are finally beginning to realize that the energy crisis is the most severe nonmilitary threat to our society in our lifetime. Unfortunately, despite this realization, we are still mired in a number of dangerous myths about the energy crisis—myths which we must dispel from our minds if we are to identify the real energy policy options and choose among them.

Those myths stem in part from our failure to appreciate the incredible complexity of the energy problem, and the fact that there are fundamental uncertainties about it that no paper studies will ever solve. This is because the energy crisis, besides being a technical and scientific problem, also presents questions of fundamental human values that are not susceptible to pat solutions.

Despite the fact that there is no simple, quick-fix cure to America's energy problems, panacea merchants abound. As part of the search for a panacea, several myths have emerged about the causes of the energy problem. These myths hamper the public policy debate, and we would all be well served if they were abandoned.

First of all, the venality of the oil companies is not a fundamental cause of the energy crisis. True, oil companies make a lot of money. People who own oil resources—whether they are oil companies or Texas farmers—are richer by far than they were a decade ago owing to the recent increase in energy prices. But the fact is that only a few cents of the cost of a gallon of gasoline can be attributed to the profits of oil companies. For example, the last time another round of fat profit figures was announced by the major oil companies, little notice was paid to the fraction of sales that were accounted for by the profits. For ARCO, Exxon and Gulf, profits were in the range of 4 to 6 percent. Most of the higher price of gasoline is going for higher prices to OPEC countries and for exploitation of more costly oil and gas reserves.

If we were somehow magically to develop a government-controlled energy sector operating at perfect efficiency and earning zero profits, it would not make much difference in the availability of energy or its price.

This is not to say that we should not tax away some of the windfall gains to owners of old oil reserves; that's a political question, with good arguments on both sides. Rising energy prices have had a major effect on income distribution, and we can expect the government to seek some way to cushion this effect, especially with respect to the elderly. At the same time, energy firms have the best information about potential new energy sources, and are likely to invest more wisely in developing energy sources than is the government.

If government wants to encourage even more research, development and exploitation relating to new energy sources, it should focus more on stabilizing energy markets and less on keeping prices down and telling energy companies the details of how to go about the job. Firms in the energy sector face two sources of risk regarding investments in energy that have little to do with the inherent promise of new technologies. One is the domestic regulation environment: will a firm be able to develop a new resource if it figures out how to produce it economically, or will regulatory policies suppress prices so that the technology is not viable, or perhaps insist that someone else be given the rights to the new source? Development of domestic oil has been retarded for these reasons, and energy companies are likely to worry that the same pattern will be repeated. The second uncertainty has to do with the international price-setting activities of OPEC. If Americans come up with a wonderfully cheap method for producing oil shale or liquids from coal, the costs will still exceed the costs of pumping oil in the Middle East. OPEC could respond by assuring that the world oil price is always just low enough to keep out the new technology.

The United States government could remove much of these uncertainties by changing its method of subsidizing new energy technology. Instead of granting new subsidies for specific research projects or to construct new facilities, as is the current practice, the government could guarantee to buy a fixed amount of each new energy source at a preannounced price whenever a company wishes to sell. This would guarantee the presence of a market, regardless of other regulatory rules or OPEC strategies.

The venality of OPEC is also not the source of the energy problem. A common opinion is that somehow monopolistic oil sheikhs from the Middle East suddenly discovered the wonders of cartelization, and thereby caused our problems by forcing huge increases in the price of oil.

Certainly OPEC has contributed to the increase in energy prices. However, right now the source of instability in the cartel is that OPEC members can receive even higher prices than are being set by the cartel, not that prices are too high.

A more important cause of increased price is the growth in energy demand. One factor is increasing industrialization of the poor countries. Another cause is rising living standards in the already developed countries. At the same time, there has been a decline in oil reserves in relation to the higher demand, particularly in the U.S. Moreover, the new oil reserves we are discovering are coming along a little less frequently and cost much more to exploit.

Another source of price increases, independent of OPEC, *per se,* is that policies of Middle Eastern countries are changing independently of OPEC. The Middle Eastern oil fields are no longer being run for the benefit of Western Europe or North America; they are being run for the benefit of the people who live in the Middle East. Oil supply and price decisions reflect rational economic calculations from the point of view of the domestic welfare of oil-producing nations. And because of their wealth, their relatively low capability to absorb investment and economic growth in their own countries, and the depletion of their oil reserves, OPEC countries, even without OPEC, would simply cut production if oil prices fell dramatically. They know the long-run value of their oil is high. If oil prices fall, their best investment opportunity is just to sit on the oil they have until the inevitable rise in prices resumes.

The real international problem is the dependence of the industrialized countries on the world oil market. The United States demonstrated its vulnerability in 1979, when the Iranian crisis led to a minor energy panic in this country. The vulnerability of the United States to instability in Iran and other Middle Eastern countries arises because of the magnitude of oil imports. Indeed, had oil prices not risen in the 1970s, oil imports would be an even larger part of energy use, and the United States would be even more vulnerable than it is today to a crisis in the Middle East.

Because we have become dependent on a handful of countries for oil, American foreign and domestic policy has been profoundly affected. Saudi Arabia stands as an obstacle to two major proposals to bring our own energy house in order. They threaten to raise their oil price if we attempt to impose an import tax to reduce our consumption of foreign oil, and they threaten to cut back their production of oil if we attempt to build up a strategic petroleum reserve to insulate the United States from another temporary crisis in the world oil market.

American relationships with Europe are also being affected by oil dependence. The United States produces over 80 percent of the energy it consumes, but Western Europe and Japan produce only about half.

Their greater dependence on oil imports pulls them away from the United States and towards the Arab world on issues involving the Middle East. And the erosion of the international position of the dollar because of our failure to control expenditures on oil weakens the rule of the United States in international economic affairs.

A third myth about the energy crisis is that the world is running out of energy. Or, in the less stark version of this myth, "The world is running out of energy except for X, so we must adopt X or die," where X is sometimes solar, sometimes synthetic fuels, sometimes coal, and even sometimes true gimmicks like gasohol.

Some believe that in ten or twenty years hence, someone is going to peer down the last oil well and observe, "That's it . . . it's all gone. Back to the Stone Age!"

Actually, the United States has plenty of energy sources available, but they cost much more to develop than the ones that we have used in the past several decades. Recoverable reserves from conventional oil and gas total two-thirds of the projected total energy consumption in the United States for the next twenty-five years. If unconventional oil and gas sources, such as oil shale, brine gas, and the like, are added in, recoverable oil and gas reserves are doubled. Recoverable coal reserves are fifteen times the total projected consumption of all energy sources in this period. Nuclear power plants could produce another half of total use by the year 2000, and if the breeder reactor were developed, a new source that is twice the size of recoverable coal reserves would be available. All of this adds up to enough energy from known sources to run the country for several hundred years with no help at all from renewable resources like solar energy.

The problem, of course, is that developing nearly all of these sources will cost more—substantially more—than the energy that we are used to buying. At prices in 1980 dollars of something around $40 to $50 per barrel, foreign oil gradually could be displaced entirely and permanently by domestic resources, even if stiff environmental controls were applied to all fuels. It is not happening because Americans do not now appear to be willing to pay this price for energy independence, and they do not want to accept the remaining environmental risks associated with producing the major substitutes that could do the job in the next decade or so: nuclear, coal and oil shale. Because we cannot decide which energy options to pursue, the vulnerability of the United States persists, and will grow worse as energy demand begins to push up against supplies of existing conventional sources. Because of the long planning horizon to construct power plants, oil shale recovery facilities, and conversion plants to transform coal into liquid fuels, the United States is facing the possibility of a major energy shortage in a decade or so, not because we lack the resources, but because we cannot make up our minds which energy development path to follow.

The fourth myth is that Americans are addicted to piggish energy use, and will overconsume it no matter what. The germ of truth to this myth is that Americans and Canadians consume more energy per capita than anyone else in the world. The other side of the story is that energy has always been and still is less expensive in the United States and Canada than in any other industrialized country. In 1975, for example, total energy use per dollar of gross domestic product (that is, the total economic production inside a country's borders) was about 50 percent higher in the United States and Canada than in Europe, and about double the figue for Japan. Meanwhile, energy prices in North America were about 60 percent of European and Japanese prices.

Do prices matter? The past decade tells us yes, prices matter a great deal in determining how much energy is consumed. For example, the government thought that Americans needed fuel economy regulation to force them to buy gas-saving autos, and so enacted a program of mandatory fleet-average fuel-economy standards. But the average fuel economy of cars actually sold in the United States since prices of fuel have been rising is actually above what were once regarded as the tough standards of the regulations. The main effect of the regulations, if any, has to be on model planning by American auto manufacturers. Without the push of the fuel standards, they might have lost even more business to foreign manufacturers.

Considering the whole energy picture, between 1972 (before the Arab oil embargo) and 1978, energy prices increased a little more than 20 percent (correcting for inflation), while the amount of energy consumed per dollar of gross domestic product declined 12 percent. Higher energy costs are causing energy to be used more efficiently, and will continue to do so as long-term capital investments are made with a view towards keeping down energy costs.

A fifth energy myth is that there is an immutable trade-off between adequate energy resources and environmental policy. Solving the energy crisis is widely believed to be possible only if we give up on the goals of clean air and clean water that were set in the 1960s. The element of truth to this myth is that a large part of the environmental problem is associated with the production and use of energy, and that each energy technology does have potentially serious environmental consequences, particularly if nothing is done to control them.

But the other important truth is that both energy and environmental policies are so far from perfection that many alternatives are available that would improve both the energy crisis and the environmental problem. More rational pricing of energy—making its price equal the true costs of production of the new resources being brought on line—would conserve energy, leading to less dependence on foreign oil and to less environmental pollution as energy production was reduced. Peak-load pricing of electricity—charging more for it in the hottest part of the summer and the darkest, coldest days of winter—would shift more electricity use to base load power plants and require less use of peaking plants. The latter are more expensive to run, and generally are more polluting. So peak-load pricing would produce lower average electricity prices and lower total emissions, again improving the energy situation and the environment.

*The final noncause of the energy crisis is the stupidity of*_____
_____. Fill in the blank with whomever you like—environmentalists, politicians, bureaucrats, technologists.

America is facing conflict over energy policy because the energy problem presents us with real conflicts and dilemmas. Different people feel different aspects of the conflict in values in different ways.

For example, the unavoidable environmental problems of some energy sources threaten values that must be protected. So, too, do rising energy prices cause real economic harm to those who have to pay them or adopt some costly conservation method. The Constitution guarantees us due process against actions that threaten our health, our livelihood, or our property rights. So it is not particularly unusual that regulatory and court proceedings will be brought to bear to resolve these conflicts.

Decisions are slow because people are more interested in being heard and in keeping government budgets down than in speedy decisions. We could allocate more money to a faster court system or a faster nuclear regulatory system, but we keep voting to reduce the size of government, not to increase it.

Our basic problem is our reluctance to face facts. Energy is more expensive and will continue to increase in price. And, we have become dependent on an unstable source of supply for a large part of our energy, namely the Middle East.

Tragically, we have not yet found the will to get out of the box in which we are trapped because we are still searching for villains—environmentalists, automobiles, oil companies, OPEC, technology—instead of looking at the real problem.

We have to decide between coal and nuclear power for electricity in the next decade or two; we have to conserve energy; we have to develop new energy sources such as solar and synthetic fuels for the year 2000, and we have to face the fact that private entrepreneurs are more likely to do this than government agencies—and to get rich in the process; and we have to increase energy prices to encourage the spread of conservation and new technology. Our future is one in which energy is more expensive, but ample to sustain our basic way of life. The sooner we face these facts, the quicker the energy crisis will end.

3

Solving the
Energy Problem

Lester C. Thurow

America's military, political, and economic positions all depend upon achieving energy independence. Militarily, the nation and its allies cannot afford to be dependent upon an unstable Persian Gulf region that sits on the borders of the Soviet Union. No matter how many ships are prepositioned, full of military supplies, the region is simply not defensible for countries that are thousands of miles away and suspect in the eyes of the local populace.

Politically the differences in the extent to which the United States and its allies are dependent upon Middle Eastern oil creates tensions in the Western Alliance that can only destroy it in the long run. A cutoff of Middle Eastern oil is an inconvenience to us but a disaster to Europe and Japan. The recent differences over Israel, boycotting the Moscow Olympics, the Iranian hostages, and the Afghanistan invasion only foreshadow even more fundamental differences yet to come. Our fundamental interests simply differ too much when they depend upon the oil that is a luxury for us but a necessity for them.

Market economies are quite good at adjusting to slowly rising relative prices for commodities (oil) that are becoming relatively less abundant, but they are terrible at adjusting to sudden price shocks, such as those in 1973-74 and 1979, or to politically inspired boycotts. And however badly the market adjusts to such shocks, the political process will make it worse. It is naive to think that individuals will sit by in a democratic society and let their real income decline without a political fight to shift the decline onto someone else. In the end the resulting political infighting may further damage the economy (witness the rules and regulations now enmeshing the petroleum industry), but this inevitable result is not going to stop the income distribution fights from starting.

The correct strategy for achieving energy independence is as clear as the need for it. Given a country blessed with abundant, if expensive, sources of alternative energy—coal, shale oil, nuclear, solar—the only problem is to bring these alternative sources on line. This requires two ingredients—time and the proper incentive structure.

Given the size of our energy imports and the need to provide alternative sources for our allies as well as ourselves, energy independence is ten to fifteen years away even if we embarked on crash programs at this moment. It will simply take that long to build the necessary mines, factories, and distribution networks. But at this moment we are still fighting about the right strategy and have not embarked on any alternative strategy. Thus a 10- to 15-year estimate is an absolute minimum.

The proper incentive structure is not to pour government money into another Manhattan project, but *to guarantee markets and prices for those that produce* alternative forms of energy. No private entrepreneur is going to invest vast sums in alternative forms of energy as long as he knows that he can always be undercut by Middle Eastern oil. A genius who finds a way to make oil for $16 per barrel will find that he has only succeeded in driving the price of OPEC oil to $15.99. The world will gain from our genius, but he will not. And since in this area genius requires an accompanying investment of billions of dollars, the genius will never emerge.

Thus it is absolutely imperative that government be involved, but not as it is now. Instead it should clearly state that it is willing to buy a certain amount of energy (perhaps an unlimited amount) at a specific price (say $50 per barrel for synthetic oil in 1980 dollars). It should then sit back and let the market respond. If after a year or two it looks like the market is not going to produce enough alternative energy, the government should raise the guaranteed price for synthetic oil for both new and old producers. This will create a greater incentive for new producers to enter the market and clearly demonstrate that there is no reason for anyone to wait for higher future oil prices. If prices go up they will get those higher prices even if they have already embarked on production. If the market is over-responding and it looks like there will be a glut of oil at the announced $50 price then government should announce a new lower guaranteed price for projects begun after that date. But projects already begun would still get the high initial guarantee price that they were promised. This reinforces the pressure to get started quickly. There is everything to gain and nothing to lose.

At the moment we have exactly the opposite incentive structure. There is everything to gain by waiting since oil prices will be going up and government will be offering greater incentives in the future. This can be seen in the coal area where everyone is buying or leasing coal fields but holding them to speculation on the future price of coal rather than to produce coal.

While the appropriate strategy is clear, the strategy has some consequences that Americans will have to learn to tolerate. Basically some Americans are going to get very rich producing alternative forms of energy while most Americans end up poorer as they pay more for their energy. And those paying more are going to have to be willing to politically tolerate those that become rich.

Unfortunately American energy consumers have only two choices. They can make some other Americans rich or they can make some middle Eastern shieks rich. And given this choice the consumer has to be mature enough to realize that making other Americans rich is better than making foreigners rich. The newly rich Americans will pay taxes, will spend their money in the United States, and support the country in political and military terms.

But even if this long-run strategy is adopted, there is that 10- to 15-year transition period. America cannot afford to wait 10 to 15 years to solve its energy problem. It is at this point that the real difficulties emerge. In the short run there is only one route to energy independence. Americans must consume much less energy than they are now consuming.

Imported oil accounts for 25 percent of our total energy consumption. But any cut of this magnitude means a drastic alteration of life styles until alternative supplies can be brought on line. The required changes will not be minor, but painful and pervasive. Such changes are not going to be brought about by voluntary conservation. We will have to force them on ourselves.

There are two possible paths. One path would be to tax oil at a very high rate so that we have no choice but to use less. President Carter's 10 cent per gallon tax and his $10 per barrel oil import tax were to have started the country down this road but they were both rejected by Congress. If America is to solve its energy problems this rejection will have to be reversed and President Carter's taxes will have to be vastly expanded.

Compare the 10 cent gasoline tax with the $2 to $3 tax that most European countries have already adopted. And most of them believe that their gasoline taxes are too small to really do the required job. At some point we will have to be willing to pay a gasoline tax of several dollars per gallon. The only question is whether we do it now and start down the road to energy independence or delay and start down the road to energy independence after some more painful experiences.

But it should be clearly understood that a large gasoline tax need not mean a large reduction in American standards of living or an expansion of government. It means a painful change in our budgets (we must buy much less energy) but it need not mean a large reduction in our other consumption. If the gasoline tax revenues are rebated to us we can use them to buy other products.

The best way to rebate the revenue from oil and gasoline taxes would be for the Federal government to return the revenue to the states on a dollar for dollar basis to the extent that they lowered their sales or property taxes. The gasoline tax would essentially replace the sales tax and the property tax with no net increase in taxation. In addition to being a rebate technique that would return the gasoline tax revenues to approximately the same income classes that contributed it, this technique has the added advantage that it would offset the inflationary impact of a large gasoline tax in the Consumer Price Index. Gasoline prices would be up but other prices would be down by an equal amount. There would be no net increase in the CPI and thus no indexed wage or prices would rise simply because the government levied an energy tax.

The only alternative to a large gasoline tax is rationing plus a white market. Here government would issue the appropriate number of ration coupons (25 percent less than current consumption) distribute them in some manner and then allow individuals to buy and sell their coupons—the white market.

There are equity advantages and disadvantages to this proposal. The coupons could be distributed in whatever manner was perceived to be equitable, but the very fact of having to distribute coupons raises the issue of what is fair. This was proven to be an almost intractable issue in the President's standby rationing program.

Consider the issue of how many coupons should be given to those in Los Angeles versus those in New York. Californians can argue that they live farther from work on average, have a poorer system of public transportation, and thus deserve more coupons per person. New Yorkers can respond that they should not be penalized for their good behavior in living close to their work and in being willing to ride and pay for a subway. Californians will never move closer to their work and learn to use public transportation if they get extra coupons. Californians can counter that in the short-run they cannot move closer to their work (it takes time to build high rise apartments) and that they cannot use public transit (it does not exist). The problem is that both sides are right and that there is no perfectly fair way to divide the coupons. Any system will to some extent be unfair and arbitrary.

Or consider the problem of how you allocate coupons among those who live in Los Angeles. Do you issue a certain number of coupons per person, per driver, per car, relative to the distance from home to work, or use some other system. Here again there is no obvious answer.

The existence of a white market makes the allocation problem a little less severe—those with special needs can buy extra coupons—but it does not obviate the problem. No one wants to buy coupons if they can persuade the government to give them more coupons. But in this case persuading the government to give you more coupons means persuading the government to give someone else fewer coupons.

Relative to the enormity of the problem the choice between taxes and rationing is not terribly important. Either system could do the necessary job, neither is perfect, and both are going to be disliked. The important thing is to pick one or the other and put it in place.

As I have just outlined, the energy problem is easily solved from the perspective of economics. The problem is only difficult from a political perspective. How are we going to be able to bring ourselves to do what we have to do? Deliberately hurting yourself in the short run to help yourself in the long run is not an American attribute. Each of us wants energy independence but each of us wants the path to energy independence to disrupt someone else's life.

That would be fine if we could find that altruistic group that would be willing to suffer all of the necessary income losses and painful changes in lifestyles. Unfortunately, no group is volunteering for that honor and our government does not have the power to force any group to endure the necessary sacrifices.

Thus, we are going to have to find a way to share the sacrifices among ourselves. If we cannot find that way, we are not going to solve the energy problem and our economy will continue the slow deterioration that we have witnessed over the last decade. As is the case with all great societies, they are never conquered from without, but are always destroyed from within. Sooner or later they face fundamental problems that they cannot solve. These problems are always technically solvable. The societies just can't politically solve them. Sometimes those fundamental problems do not emerge for thousands of years—the Roman Empire—and in other cases they emerge quickly—Alexander the Great. We may well be an example of the latter case.

4

Supply Solutions to the Energy Problem

Richard J. Stegemeier

Over 180 years ago, Thomas Malthus predicted that the world was rapidly running out of energy and natural resources. This was before oil, gas, tar sands, oil shale, electricity and nuclear power were even discovered. This persistent myth has never been stronger than it is today. But nothing could be further from the truth! We are running out of cheap energy, and of course fossil fuel is finite, but the world's fuel tank is hardly empty. Since Colonel Drake's first oil well in 1869 we have produced and consumed less than 10% of the world's known fossil fuel resources and we have barely touched the nuclear resource at all. Over 95% of the original fossil fuels in the western hemisphere are still in the ground. Let me give you a few illustrations.

Figure 4-1 shows the world's energy resources. Despite America's apparent energy problems today, the United States has approximately 30% of the world's energy resources. Significantly, the Middle East has only 1%, made up primarily of crude oil and natural gas. America's energy resources are predominantly coal, oil shale and nuclear.

In Figure 4-2 we can see the relative importance of crude oil for the remainder of this century and for the first two or three decades of the next. At the present time we have produced less than one-fourth of the expected crude oil reserves in the world. Present world production of crude oil is less than 30 billion barrels per year, and if this could be maintained for the future, then the expected 2-2.5 trillion barrels of ultimate oil recovery would easily carry us into the next century. In the meantime, enhanced oil recovery, tar sands, shale oil, oil derived from coal and other liquid fuels will come into the marketplace beginning in this decade.

Figure 4-3 describes America's fossil energy reserves in billion barrels of oil equivalent. We now produce our domestic oil at the rate of about three billion barrels per year, which would provide a 30-40 year supply based on estimates of ultimate oil reserves. Oil shale and coal are larger by several orders of magnitude than are crude oil and natural gas resources. Tar sands are not considered to be a large energy resource in the U.S. However, the Athabasca tar sands contain at least 900 billion barrels and the tar sands in the Orinoco area of South America are estimated at another 600-700 barrels.

Figure 4-4 describes the production history of America's oil industry since World War II. This oil production rate peaked at nearly 10 million barrels per day in 1970 and has had a steady decline ever since. The growing demand has been provided by crude oil imports which became approximately equivalent to domestic oil production rates in the late 1970's.

Figure 4-5 shows how America's energy imports have become heavily dependent on OPEC, which now provides nearly 84% of our crude oil imports.

Figure 4-6 is taken from a public CIA report that show the supply-demand gap which is expected to occur with OPEC oil in the mid 1980's. At the present time OPEC has excess productive capacity over the world demand for OPEC oil, but at the presently increasing rate of world demand this supply is expected to disappear during this decade. At that time there could be competition for supply with consequent increases in the price of OPEC oil.

Figure 4-7 shows the consequences of our large payments for OPEC oil. In 1977 America paid OPEC approximately 45 billion dollars for its crude oil. It is obvious that the negative balance of payments created by this is extremely important in relation to America's industrial might. It is even more significant in 1980 when the cost of imported oil could reach nearly 90 billion dollars.

FIGURE 4-1

WORLD ENERGY RESOURCES

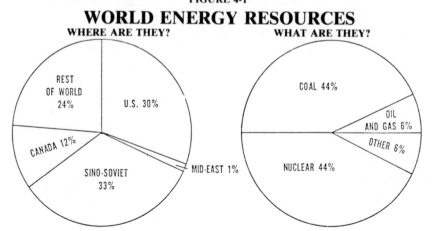

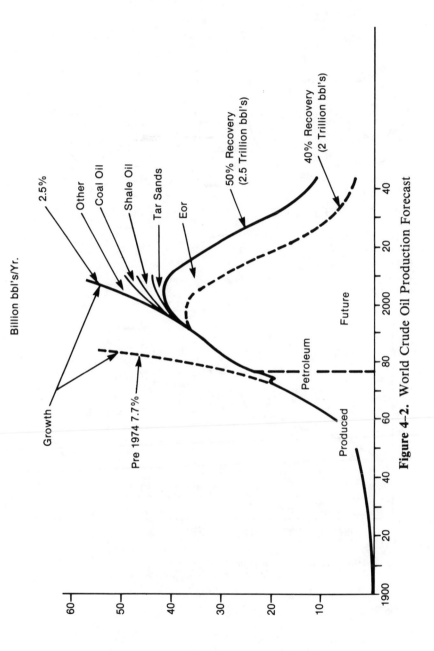

Figure 4-2. World Crude Oil Production Forecast

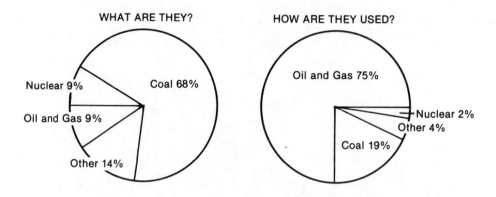

Figure 4–3. U.S. Production vs Imports

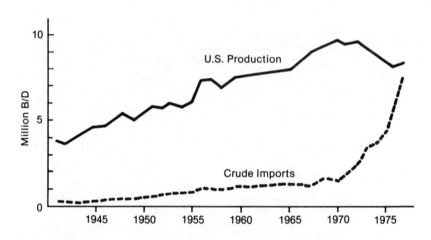

Source: *Journal of Petroleum Technology*
Figure 4–4. U.S. Energy Resources

FIGURE 4-5

UNITED STATES DEPENDENCE ON FOREIGN OIL

SOURCES OF PETROLEUM IMPORTS BY COUNTRY

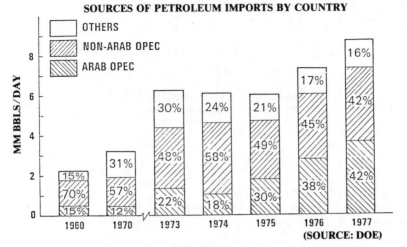

FIGURE 4-6

OPEC OIL: THE SUPPLY DEMAND GAP

SUPPLY/DEMAND MILLION BBL/DAY

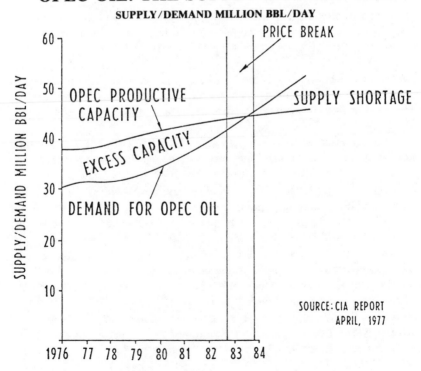

FIGURE 4-7

OPEC PURCHASING POWER
$45 BILLION PER YEAR
AMERICA'S OIL IMPORT COST
1977

THE FORTUNE 500 COMPANIES	**8 YR. 1 MO. 19 DAYS**
EXXON CORPORATION	**4 MO. 29 DAYS**
GENERAL MOTORS	**3 MO. 26 DAYS**
BANK OF AMERICA	**16 DAYS**

(Harpers, Feb., 1978)

Unfortunately, our nation has ignored the facts and has listened to the prophets of doom from the past. In the meantime, we have become captive to a foreign energy cartel through a series of bad judgments and bureaucratic nonsense. The cost of imported oil has weakened the American dollar and jeopardized our national security. Historical price controls on domestic energy have encouraged waste—not conservation. The gas-burning tiki torches in front of restaurants are a good example of our contempt for energy conservation.

We Americans represent less than 10% of the world's population—but we have almost one-third of the world's remaining fossil energy—enough to last for centuries. We have the technology and the professional expertise to develop these energy resources—but we lack the national will to make unpopular decisions.

Why, then, do we wring our hands in anguish over energy deficiency when we live in this land of abundance? The energy problem is certainly not technological—it is a political, economic and social problem. We each seek a different villain—the media, government, industry or OPEC. But the villain, if there is one, is our refusal to face the issues. Abundant, cheap energy has been supplied by the energy companies for almost a century under the system of free enterprise. It has fueled a growth of prosperity and personal freedom unmatched in the history of mankind. American energy technology has been the envy of the world. But since the 1973 Arab oil embargo, we have seen the heavy, invidious hand of governments intervene in the free market system. Where there was no shortage, one was created by government action or inaction. Let me mention just a few of these.

The 1977 budget of the newly formed DOE was almost equal to the total foreign and domestic profits of all American oil companies combined. At the same time, energy company profits were called obscene even though they were no greater than the average of non-energy companies. An entitlements program encouraged oil imports and penalized domestic production. Price controls subsidized waste when conservation was needed. Environmental and other controls slammed the doors on coal, nuclear, oil and hydroelectric development. There has been no balance between legitimate environmental concerns and economic reality.

Perhaps most deceiving has been the promise that we can somehow invent our way out of the energy dilemma in a few years—that we can find a cheap, painless, nonpolluting, do-it-yourself energy source. Never mind that these quick fixes often violate fundamental laws of science or economics, or that these idealistic solutions also have unacceptable social costs of their own. For example, with nearly 500 million people starving in the world, what is the morality of converting food products such as corn into alcohol fuel?

And finally, during the next ten years, the windfall profits (excise) tax will take away over $225 billion from essential private capital investment in energy development. All of these actions have been based on the pretense that the central planning of a federal bureaucracy is more efficient than the collective wisdom of 220 million American consumers functioning under their freedom of choice—a rather shaky premise.

The solution to the energy dilemma is clear, but it is not simple—and it is not quick. The decade of the 80s is fraught with peril for America. We could see our oil supply cut in half by political upheaval in the Middle East. Seven years have elapsed since the 1973 embargo—nearly twice the time it took to fight and win World War II. But during those seven years we have spent more time building legislative and bureaucratic roadblocks than in coming to grips with energy sufficiency.

Energy, like food, water, clothing and shelter, is a basic commodity of modern civilization. The growth of civilization has required greater use of energy to increase food production, to provide better shelter, to eliminate drudgery and to improve the quality of life. Some people will argue that we should return to a more simple life, that small is beautiful, etc.

But if we advocate a more simple lifestyle, we must define what we mean. In the 18th century, man used only 10% of the energy consumed in today's America. Nineteenth-century man consumed about one third, and Western Europe today more than half of America's present usage per capita. Even austere Russia, with few personal cars, uses almost half as much energy per capita as America.

As a citizen and member of the energy industry, I support energy conservation as the foundation of America's energy future. Conservation is the only short-term option. But can we really halve our energy use? America, because of its geographical vastness, its industrial strength, its freedom to travel, and its willingness to spend additional energy to preserve the environment, may not be able to achieve the lower energy consumption rate of, say, western Europe without a serious reduction of productivity and the quality of life.

Surely, no one would advocate reducing our farm surplus. America's ability to help feed the starving world is a direct result of our energy-intensive agriculture—from chemical fertilizers made from natural gas to mechanized farming. Our personal cars consume over 40% of our total individual energy needs and 15% of our national energy use. Ride pools and reduction of non-essential driving will have an immediate benefit. An effective public transportation system will help, but it will take at least ten years to construct. Development of a small electric car requires expansion of our electric generating capacity through coal and nuclear power. It also requires a better battery. If we refuse to permit more coal and nuclear—then forget the electric car. Smaller, fuel-efficient cars will reduce energy consumption, but at present manufacturing rates it will take ten years to replace the 110-million-car fleet in America.

In the United States we get fewer miles per barrel of crude oil because our EPA emission standards are more restrictive than in all other foreign countries. In California, we suffer another 5 to 8% reduction in automobile mileage when compared to the other 49 states because auto emission standards here are even more restrictive than the EPA. I am not taking issue with the need for pollution controls—but there is a trade-off in greater energy consumption per passenger mile when compared to other countries and other states.

A sensible energy policy must begin with aggressive conservation. In Figure 4-8, the consumption of energy in the United States is described for the various consuming sectors. Personal use constitutes about 37% of total energy consumption in the U.S. Our personal automobiles and the heating and airconditioning of our homes constitute 82% of this 37%, with the remainder being consumed in heating water, cooking, refrigeration, lighting and miscellaneous. It is obvious that the American people can have the greatest impact on conservation and reduction of energy consumption by *driving fewer miles with smaller cars* and by the prudent *utilization of heating and airconditioning in our homes.*

This alone might conceivably arrest our energy growth and carry us through the mid 80s without any additional energy requirements.

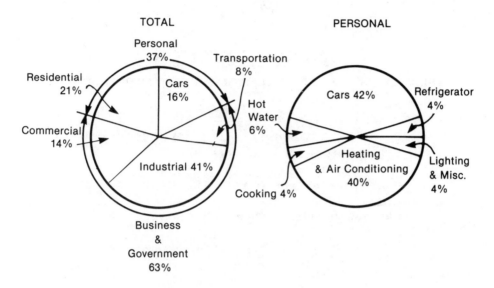

Source: Union, Science and Technology Division.

Figure 4-8. Today's Energy Use

However, by 1990 there will be another 25 million Americans in the population who will demand energy no less than the present 220 million of us. This need can only be met with growth in fossil fuel resources, coal, oil, gas, oil shale, geothermal and nuclear. So-called renewable energy has great popular and political appeal, but it will not be developed commercially on any significant scale until the next century.

To illustrate the magnitude of the problem, I would like to show you an energy equation (Figure 4-9) which identifies and quantifies the problem of energy alternatives. Let us assume that we had a goal to replace only California's offshore oil and gas production of 165,000 BPD (barrels of oil per day) with other energy resources. Remember that this 165,000 BPD is less than 1% of the daily oil consumption in the USA and only 20% of the daily oil consumption in Southern California. This 165,000 BPD is equal to the total energy output of each of the following alternative energy resources:

ENERGY EQUATION

THE TOTAL OIL & GAS PRODUCED OFFSHORE CALIFORNIA IN 1977*

EQUALS

THE ENERGY PRODUCED BY EACH OF THESE:
- **14—1000-MEGAWATT NUCLEAR PLANTS**
- **10 HOOVER DAMS**
- **500 SQUARE MILES OF SOLAR PANELS**
- **100,000 WINDMILLS EACH 100 FEET DIAMETER**
- **5000 SQUARE MILES OF CROP PRODUCTION FOR BIOMASS**

*USGS, US Bureau of Mines, AAPG.

    A.   Fourteen 1000-megawatt nuclear power plants each the size of Three Mile Island.
    B.   Ten Hoover Dams.
    C.   500 square miles of solar cells using existing technology. This is an area as large as the City of Los Angeles.
    D.   5000 square miles of wood production.
    E.   45,000 square miles of crops for alcohol production. This area is larger than the entire central valley of California.
    F.   100,000 windmills each 100 feet in diameter.

Incidentally, to replace just one 1000-megawatt nuclear power plant with a wave machine would require a mechanical device reaching from Long Beach Harbor to Monterey.

Let me also say a few words about biomass or the production of alcohols from grain and wood. I refer to Figure 4-10 which shows that even a massive agricultural effort in the United States which would double our annual grain production and double our annual U.S. wood harvest would make an almost insignificant contribution to our present production of transportation fuels. It is significant in this Figure 4-10 that the production of ethanol and methanol assumes 100% energy conversion. In fact, the energy necessary to produce, harvest, ferment and distill the various agricultural crops is nearly equal to the energy gained. In other words, without auxiliary sources of energy, such as coal, the energy economy would see no net gain from biomass production.

These are unpopular conclusions, but they are not science fiction.

The energy reality can be confirmed by any person who will take the time to make simple arithmetic calculations. Invention, no matter how great, cannot circumvent the fundamental laws of thermodynamics.

FIGURE 4-10

BIOMASS — ALCOHOLS FROM GRAIN AND WOOD

- **CONVERSION OF ENTIRE U.S. ANNUAL GRAIN CROP (OATS, CORN, BARLEY, WHEAT, ETC.) OF 9.5 BILLION BUSHELS TO ETHYL ALCOHOL WOULD BE REQUIRED TO CONVERT ALL GASOLINE TO GASOHOL (10% ETHYL ALCOHOL — 90% GASOLINE)**

- **ALL THE ANNUAL U.S. WOOD HARVEST WOULD YIELD 2 MILLION BARRELS OF METHYL ALCOHOL PER DAY — THE ENERGY EQUIVALENT TO 1 MILLION BBLS GASOLINE**

- **TOTAL CURRENT U.S. CONSUMPTION OF GASOLINE IS — 7 MILLION BARRELS PER DAY**

Conservation will have long-term benefits to our society. Also, it will give us the five years necessary to develop our own resources. *Between now and 1985 we must accelerate the development of conventional energy resources.* Oil and gas are still abundant in America. Over one million square miles of the most promising lands (one-third of continental USA) have been withdrawn from exploitation by the federal government. This is an area larger than 26 states east of the Mississippi River. Over 95% of the offshore lands (over 800,000 square miles) are also being retained by the federal government. The most promising of these federal lands should be opened for exploration immediately. Only the private sector has the technology to exploit these resources. Our oil reserves could be doubled by accelerated exploration and by the application of chemical and thermal recovery methods.

The strip mining of coal, our most abundant energy resource, should be encouraged in the western states. Coal slurry pipelines should be built to carry this coal to power-generating plants until our decrepit railroad system can be rejuvenated.

Oil shale technology is adequate to bring this resource into the system before 1984.

Geothermal development is underway and can be accelerated in the western states, especially California.

Nuclear power plant construction should be reactivated to release our precious liquid fuels for transportation.

It will not be necessary to despoil the environment with this crash program. I can think of no greater environmental horror than a military dispute over Middle East oil. Surely we can make reasonable environmental compromises. America's beauty would not be ravaged forever if a hill became a lake or if the snail darter moved upstream ten miles or if the forbish lousewort was transplanted to another meadow. (If necessary, plant it in my backyard. I've never seen a weed that wouldn't grow there!) America's common sense, which has carried us through two centuries of growth and prosperity, should not be discounted by the distortions of a few loud but impractical public figures.

In closing, let me say a few more words about energy alternatives—the so-called renewable soft energy sources. (See Figure 4-11.) These are the wind, tide, solar, biomass, etc. Despite their popular appeal, there are three good reasons why these energy resources have not been developed in the past:

1. They cannot and will not compete with conventional energy under price control. They may be free but they are not cheap.

2. They are so dispersed and at such low energy levels that enormous machines or land masses are needed to capture them. Furthermore, the energy consumed to *capture them is nearly equal to the net energy gained.* It takes only one-quarter acre of land to feed a human for a year. It would take thirty times that much land to produce enough alcohol to run his car for a year, and that ignores the energy required to grow the crop and manufacture the alcohol.

3. Even with a crash program, non fossil fuel technology will not be developed and commercialized on any large scale until the next century.

FIGURE 4-11

ENERGY OPTIONS — RENEWABLE RESOURCES

- **GEOTHERMAL**
- **HYDROELECTRIC**
- **SOLAR**
- **BIOMASS**
- **WAVES AND TIDES**
- **OCEAN CURRENTS**
- **OFFSHORE THERMAL ELECTRIC CONVERSION**
- **WIND**

The development and *replacement of a basic commodity such as energy is enormously more complex than the Manhattan Project* or the moonshot where *economics were largely ignored.* American technology can easily carry us through the remainder of this century with conventional *energy resources.* We have only a few practical options for the rest of this century. (See Figure 4-12.) I am convinced that the next century will see the development of new energy resources unknown to us today as they were unknown to Thomas Malthus in the early 1800s.

FIGURE 4-12

U.S. ENERGY OPTIONS

SHORT TERM — TO YEAR 2000
1. **REOPEN FEDERAL LANDS FOR OIL AND GAS**
2. **DECONTROL ENERGY PRICES**
3. **REMOVE DISINCENTIVES FOR COAL, NUCLEAR, OIL SHALE**
4. **CONSERVATION**

LONG TERM — NEXT CENTURY
1. **FUSION**
2. **SOLAR**
3. **SYNTHETIC FUELS FROM COAL, BIOMASS**

5 Oil Import Limit, Pivotal Move in Solution: A Citizen's View of the Energy Problem

Delmar Bunn, M.D.

The legislative limit of oil imports is not often discussed in the United States. It could, however, logically be viewed as an indispensable, central element in energy planning.

Unrestrained imports of low cost oil has been instrumental in causing high national levels of consumption,[1] retarding and dampening market signals toward timely development and production of alternate fuels and causing a hemorrhagic outflow of capital which threatens the credibility of the dollar and the economy.

The societal costs to America of importing oil in unbridled fashion have been inestimable and are increasing. The resulting negative merchandise trade balance[2] has rendered U.S. inflationary[3] and monetary problems[4] seemingly insoluble, sound economic and military systems unattainable.

It is a reasonable thesis that as surely as unchecked import of petroleum has been an essential factor in creating the nation's energy and economic problems, the restraint of oil imports is necessary to their effectual solution.

Of all major oil importing nations, this nation alone has failed to apply taxes to brake oil purchases.[5] Given America's natural and human resources, the nation could, with effective limit of oil imports, open the door to an age of plentiful energy. It is possible for the U.S. to become a net exporter of fuels and energy systems within a few years.[6]

For the decade of the 1980's and '90's, the development of alternative sources of energy could contribute to the expansion phase of a

43

long cycle in the same way that the development of electronics contributed to the expansion of the 1950s, aerospace to the expansion of the 1960's and computers to the expansion of the 1970's, as illustrated below.

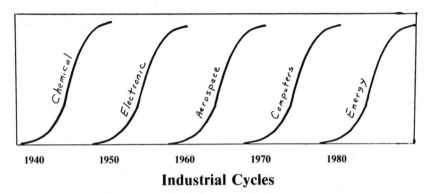

1940	1950	1960	1970	1980

Industrial Cycles

What is required? Smooth passage of the nation (as of the world) into the age of alternative fuels requires development of a formidable capital—and technology-intensive, alternate energy system.[7] The development of such a system depends in turn upon utilization of lead time and capital. Both lead time and capital will likely continue to be wasted until oil imports are impeded.

Oil is still today cheaper than alternate fuels and market forces are unlikely to reverse this for a number of years, perhaps one to two decades. This is because there is still a sizeable gap between the cost of producing oil and most of its alternatives,[8] making these decidedly unprofitable, and because it is not generally seen as in the interest of oil exporting nations to cause development of competition by changing this circumstance.

Government subsidy programs are also unlikely to change this. The magnitude of the sums involved could as well bankrupt a government as sustain its energy system. The arithmetic of subsidy indicates that it cannot be a central element in solution of the energy problem.[9]

Oil import limit may be the most central element around which policy can logically develop. Policy evolution begins on moving to the next question. How should the limit be accomplished?

The options are basically two: one, quota-rationing and two, imposition of a tax on oil prices. Quota and rationing have been amply discussed in the literature[10] and are considered desirable, if at all, only as short-term, contingency measures. Market signals are so distorted by quota-rationing that more problems are created than are solved by their use. Applied to a product as basic as energy, a plan based upon edict will threaten if not destroy a free economy.

Tax is a legitimate societal signal and tool. Price is the free market's basis. Tax can be employed to limit the import of oil into the United States without distortion if allowed to impact freely on price. What is being proposed is a two-tier tax system.

First, a non-variable, minimum-base tariff should be legislatively set. If high enough to be significant, this can deliver a powerful message to the market but also provide the assurance of price stability upon which the market can proceed with capital investments of the magnitude needed. Secondly, a variable tariff may, starting from this base, be adjusted to smoothly accomplish the objectives of the tariff.

What are some of these objectives?

First, the tariff should be adjusted to cause the merchandise trade balance to right itself.[11] The schedule by which this is caused to happen should be determined on the basis of exhibited elasticity of demand and broadest consideration of factors in the nation's economic health and strategic safety. Hence an oil import tax board.

Were the nation's foreign oil bill currently reduced by one-half, the trade balance would, of course, be strongly positive. With good energy policy such reduction could occur relatively soon.

Another objective is improved fuel utilization, which nothing motivates as well as price.

A primary goal is strategic security; decreasing dependency serves this goal without decades of expensive stockpiling.[12]

The revenues generated from an effective tariff will, further, be adequate to underwrite all governmental expenditures for energy.[13] This will make the balance of the domestic budget a realistic goal. A positive merchandise trade balance and a balanced national budget will render inflation manageable, a critical national objective.

To employ a tariff to limit the import of oil is to return to the hub of the U.S. energy problem, solving it by economic means. Present strategy of attacking energy peripherally and symptomatically at innumerable points results in ever greater governmental growth, involvement and dominance of the citizen's life and work.[14]

Solution of the energy problem can be reduced to one equation, allowing the free market to solve the problem with an efficiency which government can only emulate. Once the limit of import is economically attended, energy policy will flow forth easily, rather than being painfully and illogically developed as legislative patchwork. The nation has one decision to make: to stop the hemorrhage economically. Given the more realistic prices which a tariff will guarantee, together with a decrease in peripheral legislation, the market may then work to cure America's economic and energy woes more quickly than one now dares hope.

The need for a Department of Energy or similar agency will decrease with a rational approach to energy. That Department can be phased out in no more than ten years, likely sooner.

A SILHOUETTE OF POSSIBLE LIMITING LEGISLATION

There are a few basic points at which discussion can begin in developing legislative thought toward oil import limit by tax:

1. *The purchase of foreign oil and refinery products should be legislatively limited for each of the next ten years.*
2. *The import limit should be in dollars, not barrels.*
3. *Limit should be accomplished by a variable, board-adjusted, oil import tax, beginning from a Congressionally set, minimum-base tax, the entire tariff impacting fully on oil and refiner-product prices in free market fashion.*
4. *Revenues generated by the import tax would be employed solely for energy.*

One of innumerable possible forms of limiting legislation is as follows (numbers correspond to the above points):

1. The purchase of foreign oil and refinery products would be limited for each of the next ten years. For this period an oil import tax Board, roughly analogous to the Federal Reserve Board, with something of the latter's regional character, would propose the annual limit of that year's oil import.

   The annual expenditure for foreign oil would be proposed with, appended to, congressionally debated and approved along with the federal budget each year.

   Among the criteria which would be considered in setting the annual import limit would be the anticipated merchandise trade balance for the year, the anticipated balance of payments, the international status of the dollar and the state of the domestic economy, the implications of oil import for the GNP[15] and the provision of each region of the country with ample fuel at all times. The oil import tax board would supply Congress with the data and reasoning on which it based its proposed limit for the year.

2. The import limit would be in *dollars,* expressed in the dollar value of the federal budget.

   It is here maintained that the time for any barrel limit is long past. This is supported by the fact that imports have *decreased* 18 per cent yet expenditure for imported oil *increased* 40% between 1979 and 1980.[16]

   Any view of this nation internationally as offering a bottomless pocket for purchase of oil would be laid to rest. Only a budgeted sum would be available for purchase of non-domestic oil. The nation would make do with the amount of petroleum which that sum would buy, rather than importing to desire, then suffering the consequences of the purchase.

3. The limit of oil import would be accomplished by 1) a non-variable-base import barrel tax, together with 2) a variable tax. The latter would be adjusted to the level required to cause oil to be purchased at the rate approved by Congress for that year. The variable tariff might be set by the oil import tax board weekly.

A delay of 60 days in implementing larger tariff rate changes of over 10% would seem fair to the importer whose purchase would have left a distant port before a larger change. The adjustable tax would provide for smooth flow of petroleum into the country throughout the year, except for change in international oil prices, preventing disproportionate spending at any time of the year. It would also, because it would impact upon price, provide for availability of fuel throughout the country at all times *without* rationing and *with* competition.

Fuel costs within the nation would depend primarily upon the nation's eagerness to improve its condition. A base tariff of $10 per barrel might increase fuel prices 25ᶜ per gallon, (although due to competition and domestic production that figure could prove to be less or more than 25ᶜ). A $25 per barrel base tariff might bring fuel to near $2 per gallon (or an abundance of fuel domestically and the decreased pressure on the international market might cause a decrease in import with less upward price thrust).

Tariffs of these magnitudes would cause the nation to utilize its lead times effectively. They would also deliver powerful messages to both the domestic and world economies as to the direction of the U.S. dollar and its supporting economy.

Were the expenditure for foreign oil set at $66 billion for the next 12 months, oil would enter the U.S. at a monthly average rate of $5.5 billion. The variable tariff would be constantly adjusted to assure this.

The tax-price level would be an efficient, free-floating market mechanism accomplishing with a minimum of government involvement the goals which the nation defines as desirable. Rather than being pushed along uncontrollably, the nation would have regained control of several forces crucial to its destiny.

Tens of thousands of government regulations, large expenditures for their administration, even increasing sacrifice in national living standards and loss of economic freedom as wages and prices come to be frozen, do not all together provide the economic and strategic benefits of direct treatment of this potentially mortal lesion.

4. Revenues generated by the import tax would be used *only* for a precisely defined energy program. They would, with reduction of fuel imports, gradually decrease over the decade.

There is a strong consensus in this country that government should become smaller, not larger. Nothing of the revenues generated should be available to the General Fund or be employed to enlarge government. It is proposed that the revenues, (which with a tax of $20 per barrel, and a daily import of 5 million barrels, for example, would be $36 billion annually)[17] be used for the following four purposes:

[1] Research and development in energy and alternate fuels;

[2] Building and local assistance in building, but not operating, efficient public transportation, including mass transit;

[3] Programs improving energy utilization, including research, public education and economic assistance in, for example, decreasing auto tonnage, increasing auto efficiency, increasing building insulation and efficiency, and encouraging development of replenishable and decentralized energy sources.

[4] Support of the Department of Energy for as long as this is maintained, removing this department from the nation's general tax burdens.

There is a question among political leaders as to whether revenues can be successfully directed and kept out of the General Fund. There is some precedent for this in, for example, the use of the federal gasoline tax to build and maintain the federal highway system.

There must be a way in which, through a combination of skillful writing of legislation and close negotiation within the Congress, tariff revenues can be reserved exclusively for energy purposes and not be allowed to reach the General Fund. It is extremely important that an oil import tax serve the nation and not contribute to a cancerous growth of government.

CONTINGENCY PLANNING

A plan such as the above already constitutes a basic contingency plan. It provides the mechanisms for adjusting import of oil to any level. Once implemented, only an adjustment is required. Neither sudden, confused, expensive rationing nor a surrealistically high "disruptioи tax" of $4 or $5 per gallon would be required.[18] Instead of providing lush fields for black market and crime, immediate opportunities would be created for millions of Americans to work, to do everything from the building of windmills and the insulation of houses to the building of mass transit and the development of new sources of power.

The need for massive stockpiling, the expense of which could equal that of a war and continue for decades, is reduced to that of a modest or even a small stockpile, as dependency on foreign oil rapidly decreases.

By beginning to utilize a national characteristic, a high long-range elasticity of demand, and available lead times, each month could see the country in a better stance with respect to possible war or embargo.

Those who doubted that such elasticity exists have seen an 18% decrease in oil imports in 1980 following an approximate $12 barrel price rise on imported oil. Controlled, continuing price increases by tax can assure that this rate of decrease continues, at any rate, within reason, deemed prudent. Enactment of a system of controlled imports and a planned descent to zero dependency constitute the very best contingency planning, yet simple, efficient and lasting.

INFLATION

The ultimate net effect of almost any effective energy plan must be relatively deflationary in view of the surging inflationary forces created by inadequate policy. The inflationary effects of courageous, price-oriented, energy policy are the one-time-only, initial cost increases in fuel sensitive areas. These cause an initial, unpleasant adjustment of weeks or months which jars the economy. But this is less significant in view of the long-range deflationary effects.

Even during the "shock," "jolt" or "sacrifice" the long-term deflationary effects begin to evidence, with the strengthening of the dollar, an increased respect at home and abroad for the U. S. economy, relatively less demand for heavy metals and displacement of capital out of inflationary hedges toward productive investment. The relative effects on interest should also be strongly downward.

UNEMPLOYMENT

By swiftly reducing the outflow of American capital for foreign oil purchases and directing the large revenue immediately back into productivity, jobs will be created. Many of these jobs will demand only modest levels of skill. Training programs for these could be short.

The removal of the Department of Energy (to be supported, if at all, by the tariff revenues) and other current energy expenditures from the general tax burden, as here provided, would enable the nation to construct and maintain a sound military system. This too will create jobs. Further, an appropriate re-industrialization of the nation will naturally occur as capital begins to move out of largely unproductive inflationary hedges which decreased inflation will disenchant.

A bold but prudent energy plan can lead to near full employment. The tasks confronting the nation are enormous. America has no time for unemployment; every capable person is needed working.

INDIVIDUALS AND INDUSTRIES UNDER HARDSHIP

Initial versions of this plan contained a tapered program extending over from three to five years to provide assistance to persons and industries under hardship during the period of initial adjustment to higher fuel prices.

Since that time the windfall-profit-tax legislation has been enacted and a large fraction of the revenue to result from that tax has been earmarked for this purpose. Should that program not materialize or should the windfall-profit legislation be changed or repealed so as not to accomplish this, it would become important to develop an adequate hardship program.

POLITICAL FEASIBILITY

The denial and scapegoating of even one year ago is largely past in America. Energy has become largely a matter of public communication and education. The way is now open to national discussion and to solution of the problem.

Since time is critical, every effort must be made to assist the nation to become aware of its enormous potential in energy. This, for its own well-being, but also for the world's ultimate peace and security.

Discussion of the national energy options should be encouraged. Americans with ability to write, speak, lead discussions, and inform on an individual basis should equip themselves and become active. This is a national emergency in which citizens have a singular opportunity to serve.

As a consensus develops with respect to the direction which the nation should take in energy, corresponding adequate legislation must be developed and enacted.

What today is infeasible may be development of public consensus become completely feasible.

*   *   *

One of those alternative solutions which needs to be presented and most carefully studied is that of import limit by tax. This has the potential for changing the nation's present downhill course without delay, of constituting the direct, central, pivotal element around which successful energy policy can develop. A well-conceived program formed around this principle could reduce the nation's contingency risks in a direct, permanent manner and provide the basis for peaceful growth of American economic and political life.

Notes

1. The price of Middle Eastern oil, after two decades of prices of about $2/bbl., reached $1/bbl in late 1969. Stobaugh, Robert and Yergin, Daniel, *Energy Future* (New York: Random House, 1979) p. 25. Even spring water cost more. On the basis of such prices during two decades it was economical that anything replaceable with oil be replaced.

2. The merchandise trade deficit is developing at a rate of $40 billion annually in 1980, after deficits of $39.5 and $37 billion for the past two years. Likely deficit for the past three years: about $116 billion. "U.S. Trade Gap . . .", *The Wall Street Journal,* August 28, 1980, p. 4. Compare this 3 year net deficit with the century before the oil crisis from 1870 until 1970. The net *positive* trade balance for those 100 years, and one of the foundations of American strength, was $192 billion nominal dollars. U.S. Department of Commerce, *Historical Statistics of the United States, Colonial Times to 1970.* Part I (U.S. Gov't. Printing Office, Washington D.C. 1975) p. 884.

3. The view of inflation in America as primarily a domestic ill is here viewed as analogous to trying to stop a speeding jet with a one wheel brake. Unemployment and high interest rates do not change the international softening of the dollar associated with massive trade deficits and their effects on the dollar.

4. The Eurodollars in the world, perhaps $400 to $800 billion, far exceed M 1A. The tail has become larger than the dog. It is not possible to discuss here the integrity of viewing the Eurodollar pool as analogous to the national debt, but it must be questioned as to whether it lies in U.S. or world interests to destroy those dollars by inflation. The national debt was reduced from a sum representing 98% of the GNP (in 1940) to 38% of the GNP (in 1978) in large degree by inflation. The U.S. may have serious monetary and diplomatic problems until it takes a candid view of international economics.

5. International Monetary Fund Report. Taxes (import duties and federal taxes) on gasoline in 1979 were 300% of the U.S. rates in Japan, over 500% in Britain, West Germany and France and over 700% the U.S. rates in Italy. The U.S. tax was 14 cents per gallon.

6. This statement is based on the widest of considerations, historical and resource-wise. For many years a fully industrialized U.S. exported oil and had no shortage except in war, with a per capita domestic production of 12 barrels or less. Current domestic production is over 15 barrels (about 3.5 billion bbl annual production, 225 million people). In addition, a coal resource greater than the entire Middle Eastern oil supply, almost one-half the world's oil shale, enormous solar and many other potentials, and the capital and human resources with which to develop them, altogether make this a realistic statement.

7. An excellent discussion of this admission requirement to the new energy age is in John J. McKetta and J. D. Wall, "Energy - Process and Prospects", *World Oil,* August 15, 1980, p. 21 ff.

8. "We'll really get interested in synfuels when fuel reaches $2 per gallon," an executive of the world's largest synfuel plant construction firm said in a personal conversation in May, 1980.

9. A subsidy gap of 60ᶜ per gal. ($25 per barrel) on five million barrels per day becomes $45.6 billion annually. A figure of this magnitude would stress to the limit or destroy the domestic economy of any oil consuming nation.

10. As in Joe Cobb, "Rationing of Gas Can't Work", *Los Angeles Times,* October 13, 1980, p. 2, and Edward J. Mitchell, "Oil Import Quotas Won't Work," *The Wall Street Journal,"* Aug. 27, 1979, p. 12.

11. A projected foreign oil bill of $85 billion this year and a projected deficit of $40 billion indicates that something on the order of a 50 percent reduction in import might produce a positive trade balance.

12. A barrel of oil purchased for $33, with interest at 11% compounded annually will have cost $266 in twenty years. Site purchase and preparation, government administration, stockpile maintenance, and insurance may double this investment, bringing the cost of the security rendered by one barrel of oil over a 20-year period to over $500.

13. This plan has a self-adjusting feature. A low goal, i.e., setting the dollar import limit high, might be achieved with a $10/bbl tariff. This would supply $18 billion annually for energy and, with a relatively high import the need to create these things quickly would be reduced anyway. Should the nation elect to cut imports significantly, perhaps to $70

billion the first year, $60 the next, a tariff of $30/bbl might be required. This would make, at a 5 million bbl/day import, $55 billion available for energy purposes. Such a sum would create an energy Manhattan project two or three times over.

14. This conception of solution to the energy problem is characterized by the proposals of an economist in the department of Commerce, Dr. Michael Boretsky. In conversations in March, 1980, he advocated that *extensive, detailed regulation* become the basis of solution of the energy problem. The operation and size of motors, stoves and even toasters (yet one could, of course, buy several toasters).

15. See the chapter in this volume by James Sweeney, Ph.D.

16. Expenditure is at the rate of $85 billion in 1980, as compared with an expenditure of approximately $60 billion for 1979.

17. 5,000,000 (barrels) X 20 (dollars) X 365 (days) = $36.5 billion.

18. The disruption tax is very highly favored by the Office of Continguency Planning, Department of Energy. William Taylor, who heads that office estimates numbers of this magnitude as necessary to supplant a need for rationing with an embargo decreasing oil import by 20% or more. It is proposed that the revenue then be redistributed in some manner to the population.

6

The Relationship Between Energy Use and the Economy

James L. Sweeney

I would like to discuss the relationships between energy use and the economy. This issue has been addressed in the previous chapters, but it deserves additional attention.

There is no single objective for solving energy problems, but really a set of trade-offs among goals. First, as a fundamental objective, we must reduce vulnerability to supply disruptions, the issue addressed by Wally Baer in Chapter 12; we must realistically address energy-environment interactions while continuing to protect the environment. Third, we must deal Third, we must deal with distributional issues; most of the Congressional debate has been focused on distributional issues as a fundamental driving force to energy-policy deliberations.

Closely related to each issue is a fourth question of the overall well-being of the economy. What will the GNP be? What will be the effect of energy sector changes upon incomes or wealth of members of the economy? Those changes could be supply disruptions or increasing costs of our energy sources. This subject is what I would like to discuss: the impact on the economy of changing availability or changing costs of energy.

We can differentiate among several time horizons. First, we could examine some of the very short-run impacts. These impacts are particularly important when we consider the supply disruptions that Wally Baer discusses in Chapter 12, i.e., when, either by deliberate political action, more likely, by turmoil or war in the Middle East, a disruption in the availability of oil occurs. Associated with the supply disruptions are very rapid increases in the cost of energy. These rapid changes can significantly influence inflation and can lead to shortages. Parenthetically, most of the shortages—the gasoline lines—have been induced by regulation, by the allocation program which sends oil to where it was needed the year before the disruption, rather than to where it is needed during the disruption.

There are additional economic impacts in the short-run. Some elements of the capital stock become obsolescent. For example, the capital value of "gas-guzzling" automobiles has been reduced very significantly. Some airlines have had to scrap perfectly functional airplanes because they use too much fuel. Capital stock becoming obsolescent is a real dollars-and-cents, loss to the economy.

Structural unemployment occurs whenever the price of any factor of production increases quickly, energy included. Some people are put out of work since some activities are no longer profitable. Although these people become resources available to other sectors of the economy, there is often a time lag before the shift in employment occurs. Since some people have very specialized skills and some physical capital stock has specialized uses, the adjustment process may be slow. This slow adjustment leads to important economic losses in the short run. Furthermore, these losses would occur even in a perfectly managed economy. And we have never seen a perfectly managed economy; the U.S. economy certainly cannot be so described. . . .

Since there is only incomplete information available to macroeconomic policy makers, and only a limited number of levers to deal rapidly with sudden changes, in the case of disruptions we tend to face unemployment driven by insufficient aggregate demand for goods and services. Purchasing power is reduced; the demand for goods and services consequently goes down; additional people are put out of work; their income is reduced; and there are additional reductions in the demand for goods and services. This multiplier effect leads to additional short-run unemployment.

These are all *short-run* effects. We have seen these happenings during the '73-'74 oil embargo and after the 1979 revolution in Iran. These short-run effects are important. But they are not the issues I would like to focus on today. Primarily I plan to discuss the *longer-run* issues of economic growth with reduced availability or higher cost of energy.

There is a common belief among many governmental policy makers, people in industry, and the general public that the economy simply cannot grow without growth in the energy consumption, or to put it more precisely, that the economy cannot grow without roughly proportional growth in the quantity of energy used. This is a belief in a lock-step relationship between economic growth and energy growth.

This notion has been supported by some readily available statistics. If we plot the rate of increase in the United States use of primary energy and the rate of increase of the U.S. GNP on a year-to-year basis, we obtain Figure 6-1. The broken line represents the rate of increase of energy consumption; the solid line GNP growth. We notice that generally as the economy grew, energy use grew; as the economy declined, energy use declined. Basically energy growth and economic growth were statistically linked to one another. Such data have been used to support the lockstep theory.

If there were a lockstep between energy usage growth and the GNP growth, then restrictions on the availability of energy would be tantamount to restrictions on the ability of the economy to grow. I would like to attack the mistaken notion that there is such a lockstep relationship. First, one should note that during most of the time included in Figure 6-1 the real energy prices facing consumers were changing only slowly, as incomes increased consumers drove more and bought bigger houses. Bigger houses require more heat. More miles driven imply more gasoline consumed. Thus, consumer use of energy increased as the economy grew. In the industrial sector, if more goods and services are produced, more capital, labor and energy are used. Since energy is one of the factors of production, energy use in the industrial sector increases and decreases with the overall level of industrial production.

When prices change, however, the relative proportions of the various industrial inputs no longer remain constant and consumer purchasing power patterns change. People can adjust, substituting factors of production for one another or substituting consumption goods. How do they do this? For example, you have probably substituted some labor for energy by driving a car which is a little bit less comfortable than before. You may have linked trips together, thereby spending more time planning before you drove. Or you may have carpooled. In the industrial sector we can observe many adjustments away from energy. This is especially evident when new capital investments are undertaken. Very obvious changes are apparent in HVAC (Heating, Ventilation, and Air Conditioning) installations. Almost every major corporation now has at least one energy specialist whose job it is to find ways to use less energy. With these substitutions, which are motivated by energy price increases, output can increase while energy inputs decrease since other inputs are increasing in place of energy.

In the interest of brevity rather than going through the entire analysis, I would like to state just a few conclusions and display a few graphs that show what I believe is the relationship between use and economic growth. Those interested in the technical details should refer to the papers listed in the references.

PERCENTAGE GROWTH RATES

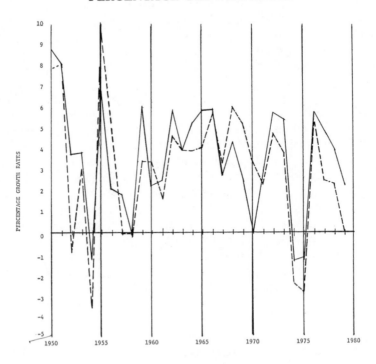

FIGURE 6-1

U.S. Energy Growth Rates versus Economic Growth Rates, 1950-1980

Underlying my analysis will be an assumption that there is some production function in the economy: with a given amount of capital, labor, and energy, there is only so much output that can be produced in the economy. If more capital, labor, or energy is available, more goods and services can be produced. The economy is always producing at its maximum for the given quantity of capital, labor, and energy utilized. Thus, implicitly, I am ignoring all regulations in the economy. (That is an heroic assumption; you will forgive me. I want some simplicity in the analysis. However, I don't really believe that our regulations are insignificant.) This set of assumptions is called the "First Best World" in economic literature. It is a world of full employment: everybody who wants to work is working and all capital is fully utilized.

Under these assumptions, my first conclusion is indicated by Figure 6-2: If the *costs* of producing energy or importing energy increase, there *will* be a significant impact on the ability of the economy to grow.

In Figure 6-2 is plotted GNP (standardized to unity in the base case) against the per-unit cost of energy (labeled as energy price in the graph), also standardized to unity in the base cases. By per-unit energy cost I mean the amount of goods and services that must be foregone in the economy in order to produce or to obtain a unit of energy. The per-unit cost of energy should be thought of as including the price of imported oil, the average cost of generating electricity, coal mining costs, and so on. Here I have assumed that the relevant energy commodities cost 6% of the GNP in the base (or reference) case.

FIGURE 6-2

**Relationship Between Energy Price and GNP
as a Function of Aggregate Elasticity
for the Cost Increase Case**

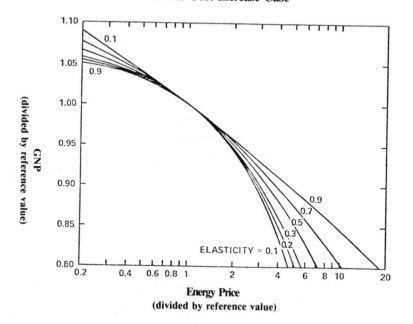

Note: Energy use and price measured at secondary level.

Source - EMF 4: "Aggregate Elasticity of Energy Demand"

Figure 6-2 shows how changing the per-unit cost of energy can be expected to change GNP from what it would be in the base case. The differences between the curves depend upon the elasticity of demand for energy. Elasticities between 0.1 and 0.9 are represented.

What do I mean by the elasticity of demand for energy? If you increase energy price, holding economic output constant, the energy demand decreases. The elasticity is the percentage demand reduction associated with a unit percentage price increase. If energy prices increase 10% and, in response, demand decreases 5%, then the elasticity of demand is 0.5. The elasticity of demand is a measure of how responsive factor proportions are to energy price changes.

Figure 6-3 illustrates how the demand for energy per unit of GNP depends upon energy price for various assumed elasticities of demand. Again, energy price and the energy use per unit of GNP are normalized to unity in the base case. Low elasticities imply that the energy use per unit of GNP will be rather insensitive to energy price; high elasticities imply the converse.

As the per-unit cost increases, the GNP decreases. And this can be a large impact. If only imported oil is included, then energy costs are 4% of GNP. A doubling of imported energy price would reduce GNP by up to 4%, depending on the demand elasticity. All oil represents almost 8% of GNP. All energy together represents more than 10% of our economy and a doubling of these costs would have a correspondingly greater impact.

Figure 6-2 illustrates an intermediate case in which the cost of the energy commodities being examined is 6% of GNP in the base case. Then if per-unit energy cost increases by a factor of, say, five, it would reduce GNP by 10% to 25%, depending upon the elasticity of demand.

Let me give you another example not illustrated by Figure 6-2. In 1979 world oil prices increased by $15.00 per barrel. The U.S. imports about 3 billion barrels per year, so that cost increase represents roughly a $45 billion annual GNP loss ($15.00 x 3 billion), two percent of our economy. The world oil price increase of 1979 roughly reduced our GNP by 2%. That is a large number!

The first proposition then is that increases in the cost of energy will reduce economic growth even if the economy is perfectly managed. If the economy is not perfectly managed, things will be even worse.

FIGURE 6-3

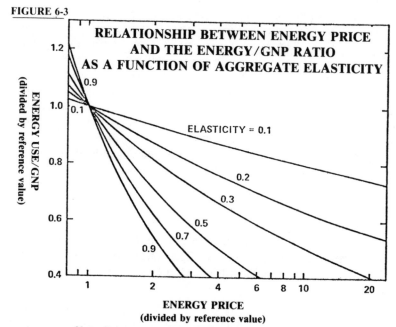

RELATIONSHIP BETWEEN ENERGY PRICE
AND THE ENERGY/GNP RATIO
AS A FUNCTION OF AGGREGATE ELASTICITY

ENERGY USE/GNP (divided by reference value)

ELASTICITY = 0.1

ENERGY PRICE
(divided by reference value)
Note: Energy use and price measured at secondary level.
Source - EMF 4: "Aggregate Elasticity of Energy Demand"

What are some examples of cost increase? A world oil price increase
is one particularly obvious example. Another stems from restrictions
against nuclear power plants and coal plants. These restrictions force
utilities to use more costly technologies for generating electricity, e.g. oil-
fired generation. Any action which forces members of the economy to
use up more finished goods and services, more labor, or more capital to
produce the same output will qualify as cost increase.

A second proposition is illustrated by Figure 6-4. Restrictions on the
availability of energy, not associated with cost increases, will have an impact on economic growth. However, this impact will be far smaller than
that associated with cost increases.

What are some examples of such restrictions? A tax on the importation of all oil or an excise tax on its use will not increase the *cost* to the
economy of importing oil but will increase the *prices* people pay.
Another instrument is price control: it leaves unchanged the *cost* of producing or importing oil but reduces the *price* that people pay. This instrument motivates members of the economy to use more energy than would
be optimal. A quota on the importation of oil would be another example
of energy use restrictions which are not associated with resource cost increases.

Figure 6-4 plots GNP against quantity of energy consumed, in the situation of energy use restrictions which are not linked to cost increases (e.g. in the case of an excise tax on all energy). The base case, against which both GNP and energy use are normalized, represents an uncontrolled, competitive market. The various curves are associated with various assumed energy demand elasticities. The curves depend greatly upon the price elasticity of demand for energy, since the parameter is a measure of how easily it is for the economy to substitute away from energy. Here again we are considering energy which in aggregate accounts for 6% of GNP (for example, this may include all imported oil and a share of domestically produced oil).

FIGURE 6-4

**RELATIONSHIP BETWEEN ENERGY USE AND GNP
AS A FUNCTION OF AGGREGATE ELASTICITY
FOR THE EXCISE TAX CASE**

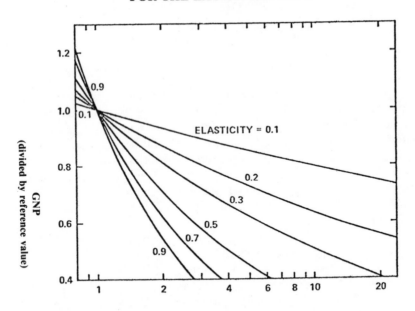

ENERGY USE
(divided by reference value)

Note: Energy use and price measured at secondary level.

Source - EMF 4: "Aggregate Elasticity of Energy Demand"

On Figure 6-4, assume that we impose a tax on energy use or impose quotas of some type in order to reduce our energy use by forty percent, to sixty percent of what it would have been otherwise. If the elasticity of demand were as low as 0.1, that policy would cost almost twenty percent of our GNP. It would be hard to justify! If the elasticity were as high as 0.9, the policy would cost one percent of the GNP. Although one percent of the U.S GNP is a large number, this policy might be sensible since our vulnerability to supply disruptions would be greatly reduced.

If the elasticity of demand for energy is large, we can gradually reduce our use of energy, while imposing only a small proportional impact on the economy.

Since the demand elasticity is so critical to the evaluation of energy use restrictions, it is useful to review the evidence. There have been many studies of demand elasticities and many models have been developed. These have been recently surveyed by the Energy Modeling Forum.* Table 6-1, deriving from that study, shows 25-year secondary aggregate energy demand elasticities. These numbers represent long-run energy demand elasticities. To obtain these responses, therefore, takes many years. The word "secondary" refers to the point of measurement. Secondary energy refers to measurement at the refinery gate or electric utility bus bar. "Aggregate" implies that an aggregate of all types of energy is considered. Elasticities are displayed for almost all of the prominent energy demand models in the United States.

Three classes of models are displayed on this table. Some models are called judgmental models. Parameters in these models are inserted to correspond to the beliefs of the model builders. Engineering models generally are based upon engineering cost studies which include representatives of many technologies and their costs. The statistical models have parameters generated through statistical inference techniques based upon historical data. These rely upon the historical record to infer the responsiveness of energy demand to prices. While most of the statistical models include just U.S. data, some include data from all the OECD countries. There are several different econometric approaches used in these models.

*EMF 4, *Aggregate Elasticity of Energy Demand,* Volume I, Stanford University, August 1980.

TABLE 6-1

25-YEAR SECONDARY DEMAND
ELASTICITIES BY SECTORS
(Paasche Index)

Sector	Statistical		Engineering		Judgmental	
Residential	Hirst Residential[a]		0.4			
	Griffin OECD	0.9	BECOM	0.6		
	MEFS	0.5				
	Pindyck	1.0				
Residential/ Commercial	Baughman-Jaskow	0.8	BECOM	0.5	EPM	0.5
	BESCOM/H-J	0.7				
	MEFS	0.5				
Commercial	MEFS	0.5	BECOM	0.3		
	Jackson Commercial[a]		0.4			
Commercial/ Industrial	Griffin OECD	0.3				
	Pindyck	0.7				
Industrial	Baughman-Joskow	0.4	ISTUM	0.2	EPM	0.7
	BESOM/H-J	0.5				
	MEFS	0.2				
Transportation[b]	BESOM/H-J	0.2			EPM	0.4
	FEA-Faucett	0.1				
	Griffin OECD	0.5				
	MEFS	0.3				
	Pindyck	0.5				
	Sweeney Auto	0.5				
	Wharton MOVE	0.2				
All Sectors	Baughman-Joskow[c]	0.6			EPM	0.6
	BESOM/H-J	0.4			ETA-MACRO	0.2
	Griffin OECD	0.5			FOSSIL1	0.1
	MEFS	0.3			FOSSIL1[d]	0.2
	Pindyck	0.7			Parikh WEM	0.1

[a]Combines both the engineering and statistical approach

[b]The FEA-Faucett, Sweeney, and Wharton MOVE results are for automobile gasoline only. These are 15-year elasticities. All runs exclude the new car fuel efficiency standards.

[c]Excludes the transportation sector

[d]FOSSIL1 Conservation

Source - EMF 4: "Aggregate Elasticity of Energy Demand"

Why do we care about the elasticity of demand incorporated in these models? One reason is that these models incorporate the best scientific evidence of the elasticity of demand.

If you examine the results from the statistical models that include all consuming sectors, you see elasticities on the order of 0.4 to 0.7. This conclusion is supported if you look at the individual sectors. I place most credence in these statistical models and in the engineering models. Thus an aggregate elasticity of between 0.4 and 0.7 is consistent with most of the empirical evidence. Elasticities of demand for individual fuels, e.g. oil, can be expected to be higher.

These empirically estimated elasticities indicate that the most relevant curves for long-run analysis in Figures 2 and 4 are those characterized by fairly high elasticities. Referring to Figure 4, we could in the long-run have a very large reduction in the availability of energy with only relatively small impact on economic growth.

Figure 4 is most relevant to the case of an excise tax since such a tax minimizes the economic losses associated with the energy use reduction. There are other mechanisms such as some mandatory conservation programs (e.g., you restrict people from buying large cars) which will have larger economic impacts. But basically, if done intelligently, we can gradually reduce the use of energy without destroying or badly wounding our economy. Large reductions in the energy growth rate can translate to small reductions in economic growth.

To examine the short-run effect of price on energy demand, it is useful to ask what has been happening recently in the United States. In Figure 6-5 is plotted the ratio of energy consumption to GNP. Before 1971 we have had an increase in the use of energy per unit of GNP (the energy/GNP ratio). However, since the oil embargo, we have had gradual but relatively steady reductions in the energy GNP ratio. This is also apparent from Figure 6-1, where the economic growth rate recently has exceeded the energy growth rate. If the rest of 1980 follows the trend set so far this year, there will be another reduction of about 2% to 4% in this ratio. The higher energy prices have been motivating a steady reduction in the energy/GNP ratio.

UNITED STATES
ENERGY CONSUMPTION (Btu) / GNP(1972$)

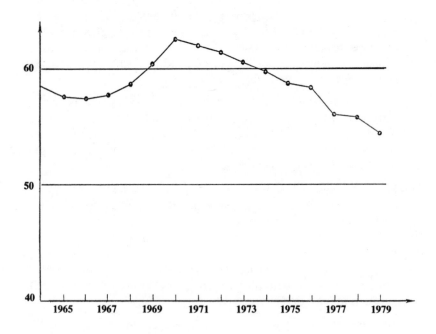

FIGURE 6-5

United States Energy/GNP Ratio, 1964-1979

I have discussed reductions in energy availability as reducing GNP. However, reductions in importation of oil could increase GNP. How? As we reduce the importation of oil, the world oil price may be reduced from what it would be otherwise. Reductions in the use of energy may reduce its per unit cost, through reductions in the world oil price. This cost reduction may more than compensate for energy availability reductions.

These ideas are illustrated in Figure 6-6. The curves which increase then decrease as a function of energy price represent the relationship between energy price and GNP if we keep the imported oil cost constant but impose (positive or negative) excise taxes. GNP is greatest with energy price being equal to the imported energy cost, GNP is decreased if we increase the energy price above or decrease the energy price below the import cost. The downward sloping line shows how increases in the energy cost reduce GNP. This line corresponds to one curve from Figure 6-2.

Assume now that we increase the domestic price by imposing excise taxes on the importation of energy. That action alone would reduce the GNP since we would be using less energy than optimal. This change alone can be represented by a movement from point A to point B on Figure 6-6. However, if we import less energy and the world oil price goes down, then we move upward along the sloping curve associated with changes in the cost of energy. This change will move us from one peaked curve to a peaked curve lying above the first. This change alone is represented by a move from point B to point C.

In this example, we started at point A and increased taxes on energy, a seemingly uneconomical policy. But because the world price of oil is reduced, we move to a higher curve, to point C. The net effect is increase in GNP. Therefore, some actions which look apparently uneconomical, e.g., some energy conservation measures, may actually *increase* GNP if they lead to reduction in the world price of oil.

FIGURE 6-6

SIMULTANEOUS CHANGES IN TARIFF AND IMPORT PRICE

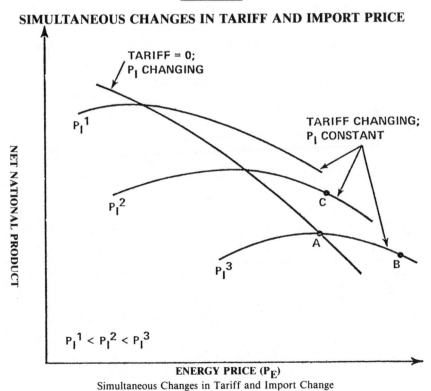

Simultaneous Changes in Tariff and Import Change

This phenomenon leads to the concept of an import premium. The import premium can justify fairly significant tariffs on the importation of oil. Depending upon the different assumptions, numbers between $5.00 and $70.00 per barrel may be the GNP - maximizing tariff on oil importation. We don't know how high is this optimal tariff, but *at least* $5.00 per barrel seems appropriate.

There is one final caveat. I have conducted most of these analyses in the so-called "First Best World" where we assume that the rate of capital formation and the rate of labor force participation are both ideal. That, unfortunately, does not describe the U.S. economy.

Another fundamental problem in our economy is related to the low rate of capital formation, one of the lowest of any industrialized economy. We have an extremely low rate of both physical capital investment and of private R&D expenditure. Partially as a result of these problems we have virtually zero productivity growth.

We heavily tax capital formation (and labor supply), even though we would like to have a greater labor force participation and more capital formation. We would like to have energy consumption and importation. But currently we subsidize energy use. Therefore, there is a great opportunity for a tax shift policy. We should place heavy taxes on the importation of energy, particularly oil, or on the consumption of oil, depending upon distributional goals. At the same time, the revenues can be used to reduce corporate and personal income taxes.

This tax shift policy would be consistent with a goal of increasing taxes on what we don't want (the importation of energy) and reducing taxes on the things that we do want (labor force participation and capital formation).

I submit that such a tax shift policy can increase our economic well being in several ways: by stimulating capital formation and labor force participation; by reducing imports of oil and therefore reducing world oil prices and reducing our vulnerability to supply disruptions.

In summary, energy *cost* increases can significantly reduce economic growth. However, reductions in energy use *can* be motivated through tax policy or other conservation programs *without* severely inhibiting economic growth. And at the same time we reduce vulnerability to oil supply disruptions, we can reduce the world oil price and reduce corporate and personal taxes which inhibit capital formation and labor force participation. There are some sensible energy policies to be undertaken. It is time to undertake them.

References

1. J. L. Sweeney, "Energy and Economic Growth: A Conceptual Framework", in *Directions in Energy Policy: A Comprehensive Approach to Energy Resource Decision-Making,* Bertram Kursunoglu and Arnold Pernmutter (eds.), Ballinger Publishing Co., Cambridge, Mass., 1979, pp. 115-140.

2. W. W. Hogan and A. S. Manne, "Energy-Economic Interactions: The Fable of the Elephant and the Rabbit?", in *Energy and the Economy,* Volume 2, EMF Report 1, September 1977, Stanford University, Stanford, California. Also in *Modeling Energy-Economy Interactions: Five Approaches,* C. Hitch (ed.).

3. EMF Report 1, *Energy and the Economy,* Volume 1, Energy Modeling Forum, Stanford University, Stanford, California, September 1977.

4. EMF Report 4, *Aggregate Elasticity of Energy Demand,* Volume 1, Energy Modeling Forum, Stanford University, Stanford, California, August 1980.

7 Energy Policy and the American Economy

Robert S. Pindyck

The odds are good that world oil prices, which rose so dramatically in the 1970's, will continue to rise over the next twenty years. Future increases in world oil prices may be moderate and gradual, the result of, say, a set of rational, economic-maximizing rules by OPEC, or increases might be sharp, sudden, and unexpected, as they were in 1974 and again in 1979. Increases in world oil prices may be moderated by new reserve discoveries and unexpected production increases by such non-OPEC countries as Mexico. On the other hand, world oil price increases might be sharply and unexpectedly accelerated by political turbulence in some of the major oil-producing countries, turbulence such as that we have already witnessed in Iran. But whatever the particular pattern is that oil prices follow, it is likely that they will rise.

This means that oil prices, and energy prices in general, will rise in the United States as well. How fast they rise will depend in part on regulatory institutions and regulatory change in this country. Because regulatory policy has held energy prices below world market levels for so many years, over the next few years energy prices in the United States will rise more rapidly than world market prices as we gradually move toward deregulation of crude oil and natural gas prices. How fast energy prices rise in later years will depend on the particular tax and regulatory policies in effect then, but at this point we can be at least fairly confident that energy prices will continue to rise, at least at a moderate rate.

What does this mean for the American economy, and what does it mean for policy—both energy policy and economic policy? Does a rising cost of energy over the next two decades imply that we will enter an era of sharply reduced economic growth, greater unemployment, and increased inflation? Does it mean a continued deterioration of our balance of payments, and further erosion of the international value of the dollar? Does it mean that the performance of the American economy will be necessarily tied to the policies and actions of a handful of oil-producing countries?

The answers to these questions depend in part on the kinds of energy policies and economic policies that we adopt in this country, and on our ability to come to grips with the facts of life of rising energy costs. One of the objectives of this chapter is to discuss the kinds of policies that are currently needed. However, before doing that, it is necessary to discuss the ways in which a rising cost of energy affects our economy.

A rising cost of energy has an impact on the economy that is both recessionary and inflationary, but it has this impact in two separate and rather distinct ways. First, rising energy prices have a direct impact on the economy by reducing the total real national income that can potentially be earned and distributed. Second, rising energy prices have an indirect impact by contributing to general inflation, and thereby destabilizing the macroeconomic equilibrium.

We discuss the mechanisms behind the direct and indirect impacts of rising energy prices in the next two sections of this paper. After that, we will be in a position to discuss the kinds of economic policies needed to respond to rising energy prices. Finally, we will conclude by discussing the implications for energy policy in the United States.

The Direct Impact of Rising Energy Prices

To the extent that energy is both an important input to production and a consumption good, with limited elasticities of substitution and demand, as it becomes more scarce the economy's production and consumption possibilities are necessarily reduced. Thus, even if an expansionary monetary and fiscal policy were successful in pushing the economy close to its full capacity level, the resulting real national income would be lower than if energy prices had not increased, and real GNP may be lower as well. It is this reduction in potential real national income (and thus potential consumption) that we speak of as the "direct" impact of rising energy prices.

It is useful to point out that the impact of rising energy prices on real national income is often confused with the impact on real Gross National Product (GNP). If the rise in energy prices occurs solely through an increase in the price of *imported* energy, real potential *output* (GNP) will remain the same, but, because of the adverse shift in the terms of trade, domestic consumption and real national income must fall. However, if the rise in energy prices occurs because of an increase in the real cost of *domestic* energy production (e.g. we exhaust cheap deposits of oil and gas and must turn to more expensive alternatives), so that more domestic resources must be used to produce the same amount of energy, real potential output must fall as well. If instead we consider *net real output,* i.e. total real output less the value of imported energy and/or domestically produced energy, this measure must indeed fall when energy prices rise. Of course in either case, the real consumption possibilities for the economy are reduced.

It is important to emphasize that this direct recessionary impact of rising energy prices is largely unavoidable, and cannot be made to disappear through the use of macroeconomic policy. As energy prices rise, the total cost of production is pushed up, and with labor and capital inputs fixed, this must mean a drop in real net output and real national income. The question, however, is to what extent must real income fall when energy prices rise? The answer depends in large part on the role of energy in industrial production. Two factors in particular will determine the impact of an energy price increase: the relative share of energy in the total cost of production, and the degree to which energy is substitutable with other inputs to production.

If energy accounts for only a small percentage share of the cost of production, even a large increase in its price will result in only a small decrease in the level of real potential income. Although the numbers vary considerably from industry to industry, for U.S. manufacturing as a whole, energy accounts for about 4 or 5 percent of the total cost of production.

If the possibilities for factor substitution are great, less expensive inputs can be used in greater quantity in place of energy. Thus, the degree to which energy is substitutable with capital, labor, and other inputs is also an important determinant of the direct impact of rising energy prices. Unfortunately, recent statistical evidence indicates that while energy and labor are indeed substitutable, energy and capital seem to be complementary, at least in the short run. In the longer run, energy and capital are also at least somewhat substitutable, but the longer run may mean something like ten or fifteen years for a turnover in the capital stock. This means that there is little room for a shift toward more capital-intensive production as energy prices increase, at least in the short run. Instead, the use of energy and capital will decrease, but with little net reduction in cost, since the substitutable alternative—labor—is already very expensive. Note that this reduction in the use of energy and capital will automatically imply a reduction in the rate of productivity growth. Rising energy prices are thus an important reason for the recently observed decline of productivity growth in the United States.[1]

Putting this altogether, in very rough terms we should therefore expect to observe a drop in real national income nearly as large as the percentage increase in the price of energy times energy's share in the cost of production. For the United States, this means that a doubling of the average price of all energy would result in about a 4 or 5 percent decrease in the real income level. The impact would be somewhat greater in Japan and the European countries, where the cost of energy is a larger share of the total cost of production.

Over the 1979 calendar year, energy prices in the United States rose by about 35 percent. Part of this was the result of OPEC price hikes, and part the result of a gradual, and long overdue, move toward energy price deregulation. However, in *real* terms this price increase was only about 25 percent. This means a *maximum* reduction in *non-energy* GNP growth of around 1.0 percent. Furthermore, this reduction could be smaller if there is room for net substitution away from energy. (And, of course, it would be smaller still if we included the profits to domestic energy producers in our real income calculation.)

This is indeed a significant and quite noticeable impact on real income, and, if the proper macroeconomic policies are not used, it would also mean a significant impact on unemployment. However, while the impact is large, it is manageable. It certainly does not imply a major economic contraction.

Another way to view the direct impact of rising energy prices is to consider the implications of gradual but continual increases in energy prices over the longer term. Although estimates of how long it will take differ, we will eventually deplete both our conventional and non-conventional hydrocarbon resources. Will this necessarily result in a catastrophic drop in U.S. income and consumption levels at some point in the future?

To take what may be the most pessimistic scenario, suppose that all of our conventional hydrocarbon reserves are depleted within the next thirty years, so that our only alternative sources are shale oil, solar, and/or nuclear power. In this case we might expect the cost of energy thirty years from now to be about five times its current cost (in real terms)—i.e. the equivalent of oil at about $100 per barrel.

Although the scenario may sound terrible, it implies an average growth rate in the real price of energy of only about 5 percent per year. Again, assuming the worst, suppose that even over the long term there is no net energy-capital substitutability. The result? Real economic growth in the United States would be reduced by about 0.2 percent per year.

This would be a noticeable reduction in economic growth, but could hardly be called catastrophic. And, this scenario ignores the possibility of technological advances that might reduce the cost of non-conventional energy supplies, or permit greater substitution away from energy, thereby reducing the drop in economic growth.

In fact, what may be a more serious problem in terms of the macro-economic effects of energy prices is the "indirect" impact. Let us now examine this second means by which rising energy prices affect our economy.

The Indirect Impact of Rising Energy Prices

Rising energy prices also contribute directly to general inflation, and, by increasing the marginal cost of production may, if wages are rigid, further reduce GNP and employment. Depending on the macroeconomic policy response to this added inflation and unemployment—i.e. whether we "accommodate" the additional inflation by using an expansionary monetary and fiscal policy to try to move back to full employment quickly, or whether we accept the additional unemployment for some time—and depending on the *effectiveness* of that policy response, there will be an added cost, namely the cost of the increased inflation and/or a further reduction in GNP. It is this secondary cost associated with increased inflation and the possible further reduction in GNP that we mean when we speak of the indirect impact of rising energy prices.

The measurement and prediction of the indirect impact of rising energy prices is much less straightforward than was the case with the direct impact, mainly because the indirect impact depends on the particular macroeconomic policy response to the energy price increase, and the design of an optimal policy response is far from clear. The intuitive inclination of some has been that the proper policy response to an energy price shock would be one of *full accommodation,* i.e. the immediate inflationary impact should be accepted, and the nominal money supply increased proportionately—or perhaps even more than proportionately to compensate for the depressing effect on investment resulting from short-run capital-energy complementarity.

This intuitive view, however, might not be correct. There are several possible reasons why it might not be desirable after all to quickly accommodate an energy price shock by expanding demand. First, and perhaps most important, there may be considerable downward rigidity in the real wage rate. There is considerable evidence that this has indeed been the case in Europe, and there is some evidence that it may also be the case in the United States. If the real wage rate is rigid downwards, a policy of accommodation will only lead to further inflation, without having any impact on employment and real output. If the real wage rate is rigid, the only policies that are useful in response to an energy price increase are those that stimulate supply.

Second, one might believe, as William Fellner has suggested, that there is an implicit "social contract" between unions, firms, and the government that hinges on a perception of the government being committed to fighting inflation. That social contract might be overturned once unions and firms saw the government "caving in" to energy-induced inflation by accommodating the energy price shock. This could result in an acceleration of inflation, and in a further institutionalization of the mechanisms (such as automatic indexing) that promote wage-price spirals. Such a concern might be particularly valid if energy price shocks occur on a regular basis.

Third, even if the real wage rate is flexible and we could ignore Fellner's "social contract" concern, full and immediate accommodation to an energy price shock might still not be desirable depending on the publicly perceived relative costs of inflation and unemployment. In a recent paper, I have shown that full and immediate accommodation to an energy price increase is only desirable if the social objective funtion is *linear* in inflation and unemployment, i.e. incremental increases in inflation or unemployment have the same perceived social cost no matter what the levels of inflation and unemployment are. If, on the other hand, the publicly perceived costs of increments in the rates of inflation and unemployment rise as those rates rise, a policy of full accommodation is not desirable.[2]

While a policy of full accommodation is probably not desirable as a response to rising energy prices, a policy of *contraction* is probably not desirable either. Unfortunately, governments often respond to the added inflation resulting from energy price increases the same way they do to ordinary inflation—with contractionary macroeconomic policies. The use of such contractionary policies can often be an unwise choice, and further contribute (unnecessarily) to the recessionary impact of the energy price increase.

To see this, consider what happened when energy prices rose dramatically in 1974. First, sharp increases in oil prices contributed directly to an increased rate of inflation. In the United States, for example, a good 3 to 4 percent of 1974's 11 percent inflation is directly attributable to the oil price increases that occurred that year. Another 1.5 to 2 percent resulted as increased demands for wheat and other agricultural exports pushed up food prices. Thus, only 5 or 6 percent of 1974's 11 percent inflation was of the demand-pull variety that we are accustomed to, and that is responsive to conventional macroeconomic policy measures. And in Japan and some of the other European countries, the inflationary impact of higher oil prices was even greater.

Unfortunately, most countries responded to this added inflation in 1974 with strongly contractionary monetary and fiscal policies. The result was the sharp recessions that we observed in 1975 in the United States, Canada, and many of the European countries. The reduction in economic growth that occurred between 1974 and 1976 was thus due in large part to our economic policies—policies that might work well against demand-pull inflation, but that unfortunately are of little use in fighting the kind of externally-induced inflation that we experienced at the time. In fact, the recessionary impact of the 1974 oil price increase need not have been as great as it was, had we adopted less contractionary economic policies.

The Implications for Economic Policy

As we explained above, the indirect impact of an energy price increase can be moderated if we avoid responding to those increases in a knee-jerk fashion with contradictionary monetary and fiscal policies. We must recognize that at least a certain amount of inflation will necessarily be associated with energy price increases, and instead of reacting to that inflation, we would be better off by keeping the economy on a steady keel.

Without seeming self-contradictory, we must, however, stress that the situation in 1979 and 1980 has been somewhat more complicated than that of 1974 and 1975, and there are a number of reasons why it probably would have been a mistake during the more recent period to pursue a policy of monetary and fiscal expansion (aimed at accommodating the increases in oil prices that occurred in 1979).

First and most important, unlike the situation in 1974, oil price increases and other "external shocks" only accounted for a small part of the startling increases in the rate of inflation that we observed in 1979 and early 1980. Inflation in 1979 was much more the result of the monetary and fiscal expansion that had come earlier. As such there was little alternative to pursuing contractionary policies.

Second, even for the *component* of 1979's inflation that could be directly attributed to oil price increases, full accommodation was probably not desirable. The main reason for this is simply that the publicly perceived costs of increments in the rates of inflation and unemployment rise as those rates rise, and rates of inflation have indeed been high over recent years.

Finally, the oil price increases of 1979 came at a time when there were other constraints on economic policy. For example, the declining international value of the dollar (brought about in part by those very oil price increases) would have been exacerbated by an expansionary monetary policy that reduced interest rates. Also, Fellner's "social contract" argument might have had particular validity in 1979 and 1980, in that accommodation at a time of unusually high inflation might have contributed to the institutionalization of fuller indexation in wage agreements.

We see then that the design of macroeconomic policy requires that we identify the inflationary and contractionary effects of the *energy price increase,* as opposed to the effects of other shocks, built-in cyclical processes, or past policies. Furthermore, the menu of policy options may be constrained by the effects of past policies, or by objectives apart from inflation and unemployment. In addition, the desirability of full or partial accommodation to energy price increases that may occur in the future will depend on the extent of real wage rigidity at the time, the availability and perceived effectiveness of "supply side" policy instruments, and of course the objective function that best reflects (again at the time) the social cost of increments in the rate of inflation and unemployment.

It is also important to recognize that to the extent that full or even partial accommodation is not the desired policy response, the indirect impact of an energy price increase may be much larger. Of course we should also recognize that there are other ways of responding to energy price increases besides the manipulation of aggregate demand. The ideal response to an energy price increase, and one with no inflationary side effect, is by matching the increase in the marginal cost of aggregate production brought about by an energy price increase with a *decrease* in that cost through "supply side" policies. Unfortunately, "supply side" instruments are limited. One possible response, however, is to reduce payroll taxes as an immediate response to any sharp increase in energy prices. Unfortunately, there are only a limited number of times in which this can be done. As a result, any "supply side" policies that are used will probably need to be supplemented with the manipulation of aggregate demand.

The Implications for Energy Policy

So far we have only discussed the impact of increases in energy prices; we have not discussed the impact of energy *shortages*. Energy shortages can indeed have very large impacts, and in general are much worse than price increases because they can result in production bottlenecks and dislocations that can seriously damage the productive capacity of the economy. In the United States we have already experienced shortages as a result of price controls, and there is always the danger that the future use of price controls might result in even more serious shortages. If this were the case, the impact of rising energy prices on the macroeconomy could be extremely serious.

As is well known, American energy policy throughout the 1970's was directed largely at keeping the price of energy low. Price controls on natural gas and crude oil artificially increased our dependence on imported oil, and thereby artificially (and unnecessarily) aggravated our balance of payments problems, and contributed to the deteriorating international value of the dollar. That, in turn, increased the prices of all imported goods, which in turn contributed to general inflation, so that Americans in fact paid part of their energy bill *indirectly* through increased general inflation.

The United States is now slowly but hesitantly beginning to turn away from these policies, and is moving toward the deregulation of energy markets. The Natural Gas Policy Act of 1978, although extending regulation to intrastate gas and creating a bureaucratic nightmare by separately regulating some twenty-odd categories of gas, at least moves us toward the eventual deregulation of most categories of gas by the middle or late 1980's. And, price controls on crude oil are scheduled to be removed completely by October of 1981.

Nonetheless, there is still a strong disagreement over energy price policies in the United States, and the future of American energy policy is not completely clear. The problem is that there is a conflict between the *economics* of energy and the *politics* of energy. The economic prescription for energy policy, with its goal of providing sufficient supplies at lowest possible costs, both today and in the future, clearly relies on the price mechanism—so that producers have the incentive to provide increased supplies of fuels, and consumers have the incentive to restrict consumption. The political prescription for energy policy, on the other hand, with its varied and sometimes conflicting goals, tries to keep the price of energy low so that the true cost of energy is hidden. Consumers then pay for their energy indirectly, through tax revenues allocated to synthetic fuel production or conservation subsidies, increased general inflation generated in part through growing OPEC imports, and sometimes outright shortages of fuels.

The first and most important objective of any new American energy policy must therefore be the complete deregulation of energy markets. This also means a commitment to refrain from regulation in the future, so that developers of new high-cost and high-risk energy supplies need not fear that their upside profit potential will be regulated or taxed away if energy prices continue to rise. The complete deregulation of energy markets is essential if we are to avert energy shortages and reduce our dependence on insecure supplies of imported oil.

However, even with deregulation our dependence on imported oil would remain at a level that is economically, politically, and strategically intolerable. The probability of a major disruption in world oil markets occurring over the next five or ten years is very high, and the United States would be much too vulnerable if such a disruption occurred. It is therefore essential that we further reduce our dependence on imported oil, even beyond what would result from price deregulation. The simplest, most effective, and least costly way to reduce this dependence still further is by imposing a *tax on gasoline*. A large gasoline tax—around a dollar a gallon—would still make gasoline in the United States cheaper than in most European countries, and would represent a clear commitment to reduced import dependence. Furthermore, the proceeds of the gasoline tax could be used to reduce payroll taxes. As mentioned earlier, this would reduce the total cost of production, and offset both the recessionary and inflationary impacts of rising energy prices.

1. For a detailed discussion of these issues, together with statistical estimates of elasticities of substitution, see R. S. Pindyck, *The Structure of World Energy Demand,* M.I.T. Press, Cambridge, Massachusetts, 1979.

2. See R. S. Pindyck, "Energy Price Increases and Macroeconomic Policy," *The Energy Journal,* October, 1980.

8

The National Energy Dividend: A Proposal for a U.S. Energy Policy

Michael D. Intriligator

1. Introduction

The National Energy Dividend (NED) is a proposal for a United States energy policy developed by the author in late 1973.[1] This paper puts the energy problem in perspective in Section 2, presents the NED proposal in Section 3, provides an economic analysis of NED in Section 4, develops a related game-theoretic analysis of the energy situation in Section 5, considers alternative potential futures for energy in Section 6, and reaches overall conclusions in Section 7.

2. The Energy Problem in Perspective

The energy problem or, perhaps, more dramatically, the "energy crisis," refers to the interrelated problems of rising energy prices, spot shortages, potential threats of disruption of energy supplies, excessive use of energy, and lack of conservation of energy. These problems emerged in the latter part of 1973, starting from the quadrupling of oil prices by the Organization of Petroleum Exporting Countries (OPEC). These problems have been made even more evident in actual or potential embargos aimed at the U.S., spot shortages, continually rising prices, and threats by certain Arab oil exporting nations to use the "energy weapon."

The United States as a nation has failed to come to grips with these problems. The later sections of this paper will discuss a policy proposed to deal with these problems, but it is important to ask why there has been a failure in past attempts to formulate and to implement policy in this area.

One reason for the failure to deal with the energy problem is the "faddish" nature of the "energy crisis." It has been hard to keep this issue at the forefront of public attention, both because of a host of other issues emerging and because the U.S. public appears to have a relatively short attention span. Other problem areas, such as poverty, pollution, nuclear energy, and many others, have each fascinated the public for various periods of time and then faded from attention despite the fact that they have not been brought to a resolution.

A related consideration is the very short-term nature of policy evaluation. The executive and legislative branches appear to have a very short horizon in terms of considering the effects of alternative policy proposals. Some issues, such as energy, however, demand a long-term horizon in view of the impacts of alternative policy options, which can extend over many years. Furthermore, government decision makers do not have tools to analyze the secondary, as well as the long-term effects of alternative policies. Finally, there are certain policy fads, just as there are certain issues that become fads. For a while there was a fad of regulation, there being a belief that regulation could solve many problems. Now the fad is just the opposite, one of deregulation, the belief now being that regulation, far from solving the problems, has in fact caused them. Of course, neither fad has much merit, both being rather simplistic ways of dealing with policy questions. Clearly there are some areas in which regulation is warranted, while others in which deregulation may be desirable.

Yet another consideration in formulating policy is the heavy reliance on technological solutions to societal problems. As a nation we are fascinated by technology and believe that it is essentially a panacea. In fact, the U.S. is, by international standards, very good at technological solutions. At the same time, however, it is not very good at political, economic, and organizational solutions to pending problems, including the energy problem.

These various considerations all apply with particular relevance to the energy problem. The U.S. public has not allowed energy to stay at the forefront of issues, despite exhortations such as President Carter's reference to the energy issue as the "moral equivalent of war." Public opinion treats the "energy crisis" as a thing of the past, rather than a current pressing problem. Policy proposals have tended to be technological and have not seriously considered long-term issues or secondary impacts, a prime example being President Carter's synfuels proposal. There has been a lack of consideration of the important political and economic aspects of energy policy in favor of technological "quick-fixes."

The rest of this chapter treats a specific proposal for dealing with the energy problem, the National Energy Dividend.

3. The National Energy Dividend (NED)

The National Energy Dividend (NED) is a proposal to deal with the energy problem in a way that will be effective over the long term and that will have desirable rather than undesirable secondary effects.[2] It avoids technological solutions, relying on economic and political initiatives.

NED calls for the imposition of substantial federal taxes or surtaxes on all final energy use, including gasoline, heating oil, electricity, and natural gas. On gasoline, for example, an additional federal tax of 80 cents to one dollar a gallon would be imposed, and comparable taxes would be imposed or added to existing taxes on other final uses of energy.

The funds collected under NED would all be placed in an energy trust fund, which would be established for this purpose. This trust fund would, in many respects, be similar to the federal Highway Trust Fund, which accumulated road user fees, in the form of federal gasoline taxes, and disbursed these funds to build the interstate highway system.

The funds collected by the energy trust fund would be disbursed in two ways. Five percent would be allocated to research and development on new and improved energy systems and on the reduction of undesirable impacts of energy use, such as environmental pollution. The remaining ninety-five percent would be returned to the public, with all adult individuals with adjusted gross income of less than $25,000 per year receiving an equal share.[3]

Only five percent is allocated to research and development since this is as much as can be absorbed by this sector. Indeed, it may exceed the absorptive capacity of the research and development sector, in which case the percentage might be reduced. Existing federal agencies, particularly the Department of Energy, the National Science Foundation, and the National Aeronautics and Space Administration, would have the responsibility for disbursing research funds to promising proposed approaches of organizations and individuals. Federal laboratories, corporations, colleges and universities, and individual researchers would all be encouraged to apply for energy research and development funding, and the most promising projects would be supported. The research and development funding provided through the energy trust fund would substantially augment funds currently devoted to this purpose and disbursed primarily by the Department of Energy.

The ninety-five percent allocated to individuals would be paid equally to all adult individuals with an adjusted gross income less than $25,000. This dividend would be provided on an equal basis regardless of energy use, location, etc. The Internal Revenue Service would disburse these funds. Those filing personal income tax returns would, if eligible on account of income, treat their NED funds as both additional income and prepaid taxes. Those not filing personal tax returns would receive a direct payment from the Internal Revenue Service, comparable to a tax refund, which could be paid as a quarterly dividend.

NED is certainly a feasible proposal. Taxes are already levied on most major energy uses, and the NED proposal would simply involve an increase in these taxes. The creation of an energy trust fund would be done via the same legislation that would create NED. Disbursements from this Fund would be managed by existing federal agencies, so no new bureaucracies need be created to implement this proposal.

NED would have several important and desirable impacts. Substantial increases in the cost of using energy would lead to a more efficient utilization of scarce energy resources, conservation of these resources, and reduction in environmental pollution. There would be substantial sums available to finance promising research and development projects. At the same time low income and middle income families would not be adversely affected since they would share in the proceeds of the new taxes. Indeed NED would generate a substantial increase in available income to low income families whose share of the dividend would be a significant proportion of income from all other sources. Finally NED would be available as a type of social insurance to all individuals. Regardless of their current level of income they will have the assurance that whatever happens to them they will receive their energy dividend. In sum, NED would

(a) ensure an efficient utilization of scarce resources
(b) finance promising energy research and development projects
(c) augment income of low income families
(d) provide an income floor for all individuals.

4. An Economic Analysis of NED

An economic analysis of the effects of the National Energy Dividend involves the study of both income and substitution effects.[4] NED involves both a change in price, leading to a substitution effect, and a change in income, leading to an income effect. In particular, under NED there would be an increase in energy prices but, at the same time, an increase in income. These two changes are illustrated in Figure 8-1. In this figure the two axes show the level of consumption of energy goods, on the horizontal axis, and the level of consumption of other (non-energy) goods, on the vertical axis.[5] The two indifference curves shown in the figure each indicate alternative combinations of energy and other goods that a representative consumer would find equally desirable. These curves thus summarize, by their slopes, the rates at which consumers would be willing to substitute other goods for energy goods. In general there is an indifference curve going though every point in the figure, with indifference curves above or to the right representing more preferred positions, more goods being preferred to less.

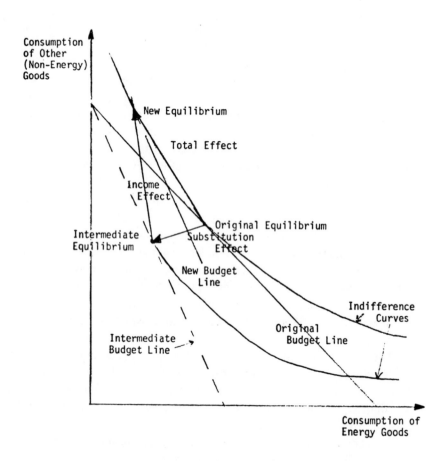

Figure 8-1: An Economic Analysis of the National Energy Dividend, using Substitution and Income Effects

The optimal budget line shows the alternative combinations of energy and other goods that the consumer could afford to buy. The original equilibrium shown in the figure is the highest indifference curve that can be reached along this budget line. It therefore shows the equilibrium consumption of energy and other goods.

Now introduce NED. Part of NED is a tax on energy goods, which has the effect of increasing the prices of such goods. As a result of this price increase the budget line shifts to the intermediate budget line, for which the maximum possible consumption of energy goods (where the budget line hits the horizontal axis) has been reduced (owing to the increase in the price of such goods) but for which the maximum possible consumption of other goods (where the budget line hits the vertical axis) has not changed (since the prices of such goods have not changed). The intermediate equilibrium is where the intermediate budget line reaches the highest possible indifference curve. The shift from the original equilibrium to the intermediate equilibrium is the substitution effect, showing the effects of an increase in price of energy goods on consumption of both energy and other goods.

The intermediate budget line and intermediate equilibrium are labelled "intermediate" because they take account of only one part of NED — the increase in prices of energy goods by higher taxes on these goods. The other part of NED is an income supplement, the energy dividend itself. The effect of increasing income, given the new prices of energy and other goods, is shown in Figure 8-1 as the shift from the intermediate budget line to the new budget line. Increasing income shifts out the budget line, but doesn't change its slope, which is determined by the prices of the goods.

The new equilibrium is where the new budget line reaches the highest indifference curve, which could be exactly the same as that reached at the original equilibrium. The income effect is the movement from the intermediate equilibrium to the new equilibrium, showing the effects of increasing income. The total effect is simply the sum of the substitution and the income effects, representing the change in consumption from the original equilibrium to the new equilibrium.

The effects of NED are thus shown as the total effect in Figure 8-1, summarizing both the energy tax and income supplement components of this plan. Note that consumption of energy goods has fallen (from the horizontal distance of the original equilibrium to that of the new equilibrium), that consumption of other goods has increased (measured by the vertical distances), and that the consumer is equally satisfied at the new equilibrium as at the original equilibrium (since both are points on the same indifference curve). In fact, when considering all consumers, rather than a single representative one, some will be better off (on a higher indifference curve) and some will be worse off (on a lower indifference curve) depending on their consumption of energy goods and income. Other things being equal, the lower the initial consumption of energy goods and the lower the initial level of income the greater the likelihood that NED will lead to reduced consumption of energy goods (promoting the conservation of energy) and increased consumption of other goods (offsetting the reduced consumption of energy goods).

The exact quantitative effects of NED on consumption of energy goods, consumption of other goods, incomes of different individuals and related matters, such as energy imports and tax revenues, depend to a significant extent on elasticities of demand for energy goods. Different investigators have reached different conclusions with regard to the effects of increased taxes largely becuase of different assumed or calculated values of these elasticities. Initial studies were highly pessimistic about the effects of such taxes because of low elasticities, but later studies were more optimistic because of higher elasticities, especially in the long run. The NED proposal, however, does not depend on any particular values for these elasticities. If the initial levels of taxes are found to be too low, then they can be later adjusted upward. In general, these taxes could be reconsidered periodically (e.g. annually) and adjusted upward or downward depending on whether the effects were considered too small or too large. A precedent is the Brazillian experience, where, after 1973, initial increases in taxes were considered too small and adjusted upward in order to attain a desired target level of oil imports.

5. A Game-Theoretic Treatment

Game theory provides a useful way of analyzing both the energy problem and the NED proposal.[6] Game theory treats various "players," who interact via competition or cooperation in a certain setting or situation—the "game"—and who receive certain "payoffs" as a result. The players and their payoffs for the energy game are shown in Figure 8-2. Four players are shown—consumers, producers, exporting countries, and government.

Players	Payoffs
Consumers	Consumers' Surplus
Producers	Excess Profits
Exporting Countries	Rents
Government	Taxes

Figure 8-2: Players and Payoffs for the Energy Game

Consumers in the context of the NED proposal refers to U.S. consumers of energy goods, although it should be realized that the rest of the world, particularly Western Europe and Japan, are major consumers. The payoff to the consumers is consumers' surplus. The concept of consumers' surplus for energy goods is illustrated in Figure 8-3. The demand curve for energy goods indicates the quantity of energy goods consumers would purchase at alternative prices. At a given price consumers will buy the quantity shown on this curve, as illustrated in the figure. The product of the price and quantity, shown as the area of the rectangle, is the expenditure on energy goods. Consumers have spent this much, and they have received this much worth of energy goods. Consumers' surplus is the net gain to consumers in these purchases over and above what they have paid. The demand curve shows how much they would have paid for each successive unit of energy goods. They would have been willing to pay a much higher price for the first unit of energy goods, shown as the height of the demand curve where it hits the vertical axis. The second unit

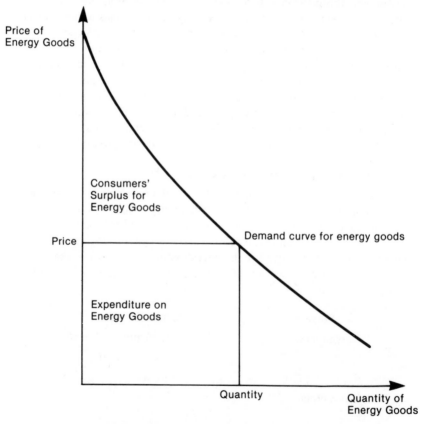

Figure 8–3. Consumers' Surplus for Energy Goods

would have commanded a somewhat lower price, but still one much higher than the actual price paid. The total of all of the amounts consumers would have been willing to pay for the quantity of energy goods purchased is the total area under the demand curve up to the quantity purchased. This amount that consumers would have been willing to pay, less the amount they actually did pay, is the consumers' surplus, shown as the upper triangular-like area in Figure 8-3. This amount, the amounts that consumers avoided paying that they would have been willing to pay for the quantity of energy goods they purchased, represents the payoff to consumers in the energy game.

The second player in the energy game is the producers, primarily, in the context of the NED proposal, the major international oil corporations, of which the largest are the "Seven Sisters." The payoff to the producers are the excess profits, by which is meant the profits earned by the producers in excess of the levels needed to enable them to provide the current and anticipated future levels of oil production. These excess profits are not identical to reported accounting profits, since they must net out normal returns to capital owners and anticipated outlays for future production. Nevertheless, excess profits and reported accounting profits tend to move together, and large values of and substantial increases in reported profits are indicative of excess profits earned by producers as their payoff in the energy game.

The third player is the exporting countries, referring primarily, in the context of the NED proposal, to the Organization of Petroleum Exporting Countries, OPEC. These countries charge various fees, royalties, taxes, etc. on the production and exports of their oil. Here their payoff is referred to as "rents" in the economic sense of this term, the payment to a factor that is fixed in supply. The oil available to the exporting countries is fixed in supply, and payments in excess of the incremental cost of producing the oil are rents. For example, it has been calculated that the marginal cost of a barrel of oil produced in Saudi Arabia is approximately 50 cents. If Saudi Arabia earns $30 a barrel on this oil then it is earning rents of $29.50 a barrel, representing its payoff in the energy game.

Finally, the last player is the government, referring primarily, in the context of the NED proposal, to the U.S. federal government. The taxes the government collects on energy goods represent the payoff to government in the energy game.

Having summarized the cast of characters, the energy game proceeds by identifying patterns of cooperation and conflict among the players in the division of the overall gains. Historically one can divide the recent era into the situation before the quadrupling of oil prices in 1973 and the situation after.

Before 1973 the major gains were the payoff to the consumers in the form of enormous consumers surplus given low prevailing prices. Producers were earning some profits, but probably not excess profits. Exporting countries were receiving some rents, but they were relatively small, as were the taxes received by government.

During 1973 there was a substantial upheaval in the division of the gains as a result of OPEC quadrupling oil prices. There were two major gainers. The first major gainer was the exporting countries, who received enormously higher rents, the petrodollars earned by OPEC. The other major gainer was the producers, who received enormously higher profits, indicative of substantial excess profits. The major loser was consumers, who lost their consumers' surplus as prices rose rapidly (note in Figure 3 that as prices rise quantity falls and consumers' surplus also falls). In effect, perhaps implicitly, the exporting countries and producers agreed to form a coalition so as to increase each of their payoffs at the expense of consumers.

The NED proposal can be considered, in this game-theoretic analysis, as one in which there is a coalition between consumers and government. By having the government tax energy goods and provide income supplements to consumers with the proceeds, consumers could recoup what they lose in consumers' surplus via the income supplements. If the 1973 changes could be considered formation of a coalition of exporting countries and producers, the NED proposal would represent the formation of a counter coalition of consumers and government, to offset the considerable power of the other two players. There would be a situation of countervailing power, the power of one side, the exporting countries and producers, being offset by that of the other, the consumers and government.

The game-theoretic treatment of the energy problem could also lead to consideration of other possible coalitions among two or even three of the players in the game. For example, the producers-government coalition would lead, as one possible initiative, to decontrol of oil prices, an action taken by government which would lead to substantial increases in excess profits of producers. Government proposals of excess profits taxes on the producers simply represent a way of dividing the gains between members of the coalition. Again the main loser would be consumers in the form of higher prices and reduced consumers' surplus. Other possible coalitions can also be treated using this analysis.

6. Alternative Potential Futures for Energy

Alternative potential futures for energy can be considered in the light of the previous analysis, particularly the game-theoretic analysis of the energy problem.

One potential future is a continuation of the status quo, involving further price increases by OPEC, with the potential for threats and political influence by members of OPEC, individually and collectively.

There is also the potential for short-term disruptions of oil supply due to instability in the oil producing regions and accidents, e.g. the sinking of an oil tanker in the Strait of Hormuz, the narrow inlet to the Persian Gulf, that could almost literally bottle up world oil supplies.

Another potential future involves the NED proposal, as already discussed, involving a coalition of consumers and government to rival the coalition of exporting countries and producers and to shift the gains away from OPEC and the Seven Sisters in favor of consumers and the government.

A third potential future involves the possible evolution of OPEC. One possible evolution would be an expansion of OPEC, via the inclusion of oil exporting nations that are not now members, such as Mexico and China, enhancing the power of and the payoff to OPEC. Another possible evolution would be the gradual breakup of OPEC due to internal disagreements, particularly those between Saudi Arabia and the other members. Saudi Arabia might find that it is in its interest to act alone. It might even be in its interests to destroy the remaining part of OPEC, which could, if acting in unison, rival the economic and political power exerted by Saudi Arabia. In this situation one of the players in the game that had been acting as a single player would be broken into several pieces, as the collusion of the exporting countries in OPEC fails. From the game-theoretic analysis one would expect, as a result, that the total payoffs to the exporting countries would diminish.

A fourth potential future involves collusion among the governments of importing countries, particularly the U.S., Western Europe, and Japan, in the form of a coalition that would rival OPEC and which might be labelled OPIC, the Organization of Petroleum *Importing* Countries. If such an entity emerged it could engage in bilateral negotiations with OPEC in much the same way that labor unions and corporate management negotiate employment contracts, setting oil prices, quantities, etc. From the game-theoretic analysis such a development would result in gains to consumers and government at the expense of exporting countries and producers. In fact, such a coalition could be considered the international analog of the domestic coalition envisaged in the NED proposal, involving the formation of a counter coalition to offset the power of a coalition among the players in the game.

A fifth potential future would add another player to the energy game, namely the Soviet Union and the Eastern bloc. So far the Soviet Union and its allies have not been involved in the international energy game, but in the future they will very likely become involved, as their demand for oil outstrips available supplies. The Soviet Union is currently the world's largest oil producer, but, as it increases its demand for oil and as its Eastern Europe allies increase theirs they will have to look elsewhere for added supplies. The Soviet Union has already started pilot coal liquefaction plants, but they will probably be inadequate. If the Soviet Union has to go elsewhere for oil the logical place for them to seek it is the Middle East. In recent years the Soviet Union has been positioning to put itself in a strong geopolitical situation vis-a-vis the Middle East oilfields. They border Iran; they have close ties to Syria and Iraq; they have influence in Ethiopia and Yemen; and they have invaded Afghanistan. Thus they have, in essence, totally encircled the Middle East oilfields. While they may not be planning a direct invasion of the producing states, they can, again from a game-theoretic standpoint, significantly threaten these states and, as a result, negotiate favorable terms.[7] The upshot of the Soviet Union as an added player would be a considerable gain to this new player at the expense of all the current players, including consumers in the West, OPEC, and Western oil producers.

7. Conclusions

While the potential futures for energy look grim indeed, the proposal for a National Energy Dividend, involving substantial taxes on energy goods with the proceeds used primarily for direct payments to individuals and secondarily for support of energy research and development, is a hopeful one. It could go far to counter the coalition of oil exporting countries and oil producers, particularly if coupled at the international level with the formation of OPIC — the Organization of Petroleum Importing Countries. There is a real question, however, as to whether the U.S. has the ability politically to implement such a proposal in view of the adverse political implications for an Administration and legislators supporting taxes which would increase energy prices. There is similarly a real question as to whether the U.S. has the influence and ability internationally to help organize a coalition of oil importing countries.

Footnotes

1. For previous discussions of the National Energy Dividend see Intriligator (1974, 1975, 1977).
2. See the references cited in the past footnote.
3. For a theoretical analysis of a related approach to income redistribution see Intriligator (1979).
4. For discussions of income and substitution effects see Intriligator (1971, 1978).
5. Of course all goods involve use of energy directly or indirectly, but some are more energy intensive than others. Thus "non-energy goods" are those that are least energy intensive.
6. For a discussion of game theory see Intriligator (1971).
7. For a theoretical analysis of the influence of threats on final payoffs see Brito, Buoncristiani, and Intriligator (1977); for an application to the distribution of world wealth see Brito and Intriligator (1978).

References

Brito, D. L., A. M. Buoncristiani, and M.D. Intriligator (1977), "A New Approach to the Nash Bargaining Problem," *Econometrica,* 45: 1163-72.

_____ and M. D. Intriligator (1978), "International Power and the Distribution of World Wealth," in Nake Kamrany, Ed., *The New Economics of the Less Developed Countries,* Boulder: Westview Press.

Intriligator, M. D. (1971), *Mathematical Optimization and Economic Theory,* Englewood Cliffs, N.J.: Prentice-Hall, Inc.

_____ (1975), "Statement" and accompanying letters. *The Energy Crisis and Proposal Solutions, Panel Discussions before the Committee on Ways and Means, House of Representatives,* Ninety-Fourth Congress, Part 2 of 4, March 6, 7, 1975, pp. 888-90, Washington, D.C.: U.S. Government Printing Office.

_____ (1977), letter to John M. Martin, Jr., *Tax Aspects of President Carter's Energy Program, Hearing before the Committee on Ways and Means, House of Representatives,* Ninety-Fifth Congress, Part 3 of 3, May 26, June 1, 3, 6, 1977, Serial 95-19, pp.294-5. Washington, D.C.: U.S. Government Printing Office.

_____ (1978), *"Econometric Models, Techniques, and Applications,* Englewood Cliffs, NJ: Prentice-Hall, Inc. and Amsterdam: North-Holland Publishing Co."

_____ (1979), "Income Redistribution: A Probabilistic Approach," *American Economic Review,* 69: 97-105.

9

A Comparison of Major Studies of Energy Policy

Eileen Alannah Orrison and
Nake M. Kamrany

During the past few years, several study groups have attempted panoramic analyses of the energy crisis. All these studies concentrate on the same fundamental issues, but do show significant differences in their policy recommendations. There are first the issues of policy tools to be used, and second issues of what are the underlying problems: issues, that is, of the "world view." It is unquestionably true that there was much more consensus on the latter group.

Policy issues include the use of market forces to provide information and efficiently allocate energy resources, fostering R&D (especially in the nuclear, conservation, and renewable-resource technologies), oil stockpiling to weather supply shocks, and expanding government regulation and standards in the private sector to more directly implement desired activities.

The underlying problems included environmental and health issues, threats to national security and economic growth, and the distributional effects of market policies, especially those of price decontrol.

Six important works are briefly reviewed:

First, the Ford Foundation Report, *Energy: The Next Twenty Years,* was composed in 1979 by a study group headed by Hans Lansberg. This is the most comprehensive of the studies.

Second, we considered *Energy in America's Future,* by Resources for the Future. Heading the group is Sam Schurr. This report, from 1979, is also comprehensive, but stops short of specific policies for all major issues.

Third, the American Assembly of Columbia University, in a project designed by John C. Sawhill, produced *Energy Conservation and Public Policy,* also in 1979.

Fourth, the report of the Energy Policy Project at the Harvard Business School, *Energy Future,* written in the same year. Fifth, the National Academy of Sciences, and sixth, the Department of Energy, are currently releasing reports (1980). The NAS report stresses technological changes, while the DOE report addresses most strongly issues of national security as imperiled by reliance on imported oil. Hence, these are the most restrictive in scope of the works considered.

FORD FOUNDATION REPORT

The Landsberg report is at once the most comprehensive in scope and most *pro-market in policy orientation.* Virtually all of the recommendations hinge upon price reform, as the crisis itself is here seen as essentially *a cost problem,* specifically that cheap energy is more scarce and abundant energy more costly than previously believed, so that *adjustments to* the true situation must be made. The problem then becomes one of *managing* the necessary adjustment so that the least amount of disruption takes place.

The general policy prescription is that the price of energy should reflect its *replacement costs* plus the imputed costs of externalities. Thus the call for decontrol of oil and gas producer prices and the reform of utility pricing. Even the recommendation that government aid R&D is coupled with the proviso that the deployment of new technology and the management decisions *must come from within the private sector.* And within this context, then, price reform becomes necessary to efficient decision making in R&D. And it is important to note that one of the areas in which R&D must be expanded according to the report is in *non-hardware research:* i.e., in those areas where *institutional* and *market forces* might be better used to accomplish goals such as *increased conservation.*

Even the environmental impact of energy use is to be subjected *to cost-benefit analyses* on a situation-by-situation basis rather than be subjected to regulation such as "no damage" levels, which the authors consider to be a scientifically untenable position.

The encouragement of conservation and the development of solar and other nonconventional technologies, although they may require some tax-subsidy manipulations where externalities are found to be present, are also dependent upon price reform, so that they are correctly priced in relation to fossil fuel alternatives. It is felt that the controlled prices in fossil fuel markets lead to overuse of those resources and discrimination against solar and other technologies. That is, although solar energy may truly be more costly than, for example, oil use in some application, the controlled price of oil makes the solar alternative appear relatively more expensive than it really is.

The most active and interfering role the government is to have under these proposals is in the area of research and *development in the nuclear industry* and in coal, again subject ultimately in most cases to *private-sector decision.* (A connected role is in increased public information about these alternatives.) The Foundation reports exceptions to the aversion to governmental R&D are in the areas of *breeder-reactor development,* coal utilization technology, some synfuel development, and research into the climatic impacts of increased carbon dioxide from expanded use of coal and oil.

There is also some moderation of the market-first stance in conservation and solar energy. There should be, argues the report, a role for *subsidy, investment mechanisms,* and some *"proving-ground" tactics* on the part of the government. Yet this manipulative framework is seen as feasible only upon the basis of accurate information about alternatives, as reflected in free market prices.

RESOURCES FOR THE FUTURE (RFF)

The RFF report offers also a series of recommendations *emphasizing pricing* policies. It stresses to a greater extent some *political issues, but seeks to solve distributional conflicts through general macroeconomic policy.* This frees each specific energy policy from responsibility in this problem.

The report is special for its specific analysis of *conservation potential* in automotive transport, residential heating, and industrial cogeneration. It also has a strong emphasis on conservation as essential for the emergence of the policy consensus the authors feel possible. This "consensus" is the true issue here: is there a general set of goals and policies that can be widely agreed upon, and, hence, politically feasible? Energy policy must allow economic growth to continue for such a consensus to peacefully emerge, so that areas in which less energy can be used without affecting output must be fully exploited.

Specifically, RFF urges not only gasoline price reform but also mileage regulation and standards, public information and persuasion efforts, and active government R&D in the area of automotive transport conservation. In residential heating, the study recommends in addition to replacement-cost-plus-externalities pricing of fuels an increased role for R&D and demonstration, including such tools as performance indices. These is also a call for government procurement of conservation technology and some micro-level regulation: thermal standards for new construction and increased utility involvement in conservation investments.

Industrial cogeneration is troubled by institutional as well as technological and pricing difficulties. Therefore, policy prescriptions must include reform of utility regulation to encourage this type of production/conservation and allow for the entry of firms using cogeneration into the energy production/marketing industries. Cogeneration requires thus a re-examination of the "natural monopoly" ascribed to the electricity utility, as generation if not transmission of electricity can become more competitive as more firms practice cogeneration.

Yet the report hesitates to be so specific about policies for synfuel development. It offers two possible scenarios: the first is an "information and insurance" program, where the government provides some assistance but refrains from active development. Here, the effort is to learn whether it is technologically and economically feasible to rely in the future substantially on synthetics, and to reach the stage where a supply shock in the international oil market may be alleviated by the bringing online of a substantial number of synfuel plants.

The second, more active, approach is to encourage and facilitate commercialization. Here, the policy is to interfere by participating in development, in such ways as by improving the synfuel industry's access to raw materials on government lands.

RFF points out some of the costs and benefits of each approach, but neglects to support one or the other. It is merely noted that "too-early" commercialization would lead to unnecessarily high costs, while too much caution might have not only economic but also political and strategic consequences if the international oil situation worsens.

For nuclear energy, the report urges increased R&D, along with limited demonstration and preparation for the breeder reactor, and better public information on the comparative safety and environmental problems of the nuclear vs. other energy industries. The study calls also upon the government to reduce costly uncertainty in the nuclear industry by accepting a large measure of the financial risk in R&D. Here also is a call for international cooperation in research.

COLUMBIA UNIVERSITY

The Columbia study had a tighter focus: conservation. Nevertheless, it expressed much the same world view as the previous studies: (1) costliness and uncertainty of supply, with, however, explicit recognition of resource exhaustibility, which the other studies lacked; (2) capital-intensiveness and long lead times of domestic energy alternatives; (3) the pervasiveness of economical problems of energy use and (4) the possibilities for continued economic growth despite reduction in energy consumption.

As with all the other studies, price reform is viewed as an important tool, and the authors analyze three policy options: (1) allowing energy prices to rise to world levels, (2) allowing them to reflect replacement costs, and (3) enforcing price increases or taxation beyond the above to stimulate conservation or finance cleanup of some energy-producing procedures. The report finds options (1) and (2) especially compelling and from its own study of energy/economic growth relationships comes to the conclusion that relatively rapid price movements can be tolerated without excessive hardship on the economy.

Manipulative taxes and subsidies were also recommended to favor conservation, materials recycling, and renewable resource utilization. There is a stronger statement here also for standards and regulation to decrease uncertainties in these newer more unconventional industries.

For conservation, the report is optimistic. It suggests that special price manipulations (such as peak-load pricing), organizational change in the conservation industries, and reform of nonhardware regulations can all significantly increase energy efficiency.

The Columbia study is also supportive of mandatory standards for mileage, construction and insulation, and calls for the government to provide better consumer information. But again, there is a wariness about excessive government interference and the "self-defeating" creation of "energy bureaucracies." As with the Ford Foundation and Resources for the Future, the first and strongest recourse is to the market, with regulatory policies to be used only when markets fail.

HARVARD UNIVERSITY

The Harvard Business School Energy Project's study has several marks of distinction. It is the most willing to advocate playing with the market, using manipulative taxes and subsidies to encourage the development of conservation and solar technologies. It has a more moderate stance on the decontrol of oil and gas prices, as well.

The Project recognizes the political difficulties of the price increases that would follow decontrol. These come from the struggles among energy producers, energy consumers, and the federal government surrounding the distribution of the "windfall" value increase of proved U.S. reserves and of the incidence of increased energy prices. Because of these struggles and also because of equity and impact considerations, the only politically tenable target for oil prices, the report suggests, is the world level. Even this target should be only gradually approached, and the costs of externalities can not be included.

However, this adds inducement to the above taxes and subsidies for nonconventional technologies. Without replacement-cost-plus-externalities pricing the system suffers inefficiency, so the authors hold that it cannot be shown that manipulations are more damaging than in action in these markets.

The Harvard report is the most supportive of renewable resource energy and conservation, claiming, especially for solar, that positive externalities do exist. This stands in direct contrast to the Ford Foundation and RFF reports, which rely more on coal, nuclear and synfuel development and rarely say more than that possible externalities in renewable resource energy should be "investigated" and those energy sources subsidized if such externalities are found. The Harvard report is also more optimistic about the short-term economical production of these energy alternatives, and about the degree of participation they can play in energy industries. This, however, necessitates active government participation in research, development, demonstration *and* deployment of the nonconventional technologies. There is much less here of the aversion the other studies show to increased government participation in R&D and regulation.

DEPARTMENT OF ENERGY

The Department of Energy report addresses the aspect of the energy problem that is the potential disruptive effect of international supply shocks due to political instability and military aggression in oil-rich areas. Here, the emphasis switches from the energy crisis as a cost and adjustment problem to the crisis as a strategic and tactical problem.

The report provides three different strategies, tied to three different perceptions of danger. The special tool is oil stockpiling, and the issues are those surrounding the correct rate of accumulation within the stockpile. Nevertheless, they too stress price reform as essential to a solution of the energy problem.

The study presents alternatives following two classifications: economic policy and energy policy.

Economic policy addresses itself to the inflation and employment changes exacerbated by supply shocks and oil price increases in the world market. The difficulty centers on the stagflationary aspects. The procedure recommended is the construction of welfare loss functions with inflation and GNP losses as arguments, and then choosing expansionary or anti-inflationary policies based upon the impacts on these functions.

Specific energy policies are treated separately. A breakdown of policies into four categories is made: (1) "short run contingency plans," (2) "oil import reductions," (3) "lead time and uncertainty reductions," and (4) "increased and diversified foreign supplies." All these policies address the related problems of disruption, high and uncertain oil prices, and military and strategic issues. The report offers specific quantitative recommendations, depending upon the world view of the international oil system's probabilities of disruption.

NATIONAL ACADEMY OF SCIENCES

The NAS shares the same general world view of all the studies: It, too, notes no intractable energy/economic growth linkage. In fact, it is decidedly optimistic on the possibilities of the technological flexibility. The report further embraces price reform as a necessary step in any energy policy. Moreover, it makes several highly specific technology-related policy prescriptions, among which are (1) continue concurrent development and use of coal and nuclear energy, (2) put a high priority on developing synthetic fuels from coal and oil shale, and (3) increase research in non-conventional energy sources, with special attention to energy storage. The special status of storage is that many advances in this field are not source-specific; that is, they can be used for any number of renewable resource technologies.

The only difference here between the NAS and the previous studies is its break with the Harvard and Columbia works on the short-run applicability of non-conventional sources, on which the NAS report is decidedly more pessimistic.

The NAS report is prepared by the Committee on Nuclear and Alternative Energy Systems (CONAES). The focus is on the years 1985-2010, seen as the transition period: the transition from depletion resources of oil and gas by well-developed technologies to using new technologies of high and uncertain costs.

The report finds conservation, especially as encouraged by price increases, caused by decontrol and special use disincentives in the imported oil market, essential to a smooth transitional period. The report also notes that conservation is the least risky of all energy actional alternatives. Most substantial conservation technologies, it notes, are available only with time for a new generation of capital, and it is underscored that efficient choice of this capital requires correct pricing.

The conservation needs to be supplemented by increased domestic oil and gas production by new recovery methods, and especially by synfuel production. Again, coal and nuclear power are to be heavily relied upon in this period.

The transition must also, however, be a time of research into the more permanent alternatives of renewable resources, possibly including fusion, and definitely including the breeder reactor. This reactor is viewed by the NAS as essential if the maximum projected rate of energy demand is experienced in the future.

Thus the report centers more on what the DOE calls "energy" rather than "economic" policies; the report is a collection of specific recommendations on the necessary direction for energy technology development.

SUMMARY

All studies, then, agreed on the following points:

(1) The energy problem is one essentially of adjustment to changing and unstable economic and strategic costs; it is not a problem of absolute physical scarcity.

(2) Much flexibility is possible in the relationship between energy and economic growth, thus allowing for better chances of smoothness in the needed adjustments.

(3) Research and development deserves governmental support to at least some extent in more energy industries, because of the large financial risks and possible positive externalities involved in some energy alternatives.

(4) Finally, the most prevalent, and most frequently and eloquently argued policy prescription is some kind of pricing reform. Whether distributional issues are addressed, and whether externalities are to be considered as well, there seems to be ubiquitious agreement on the necessity to increase oil, gas and utility prices to at least near world levels, so that the information on scarcity as well as the desire to conserve would be increased for the consumer. This would accomplish the essential target of all reports: increased rationality and efficiency in energy/economic decision-making.

10 An Economic Profile of Major Presidential and Congressional Initiatives to Deal with the Energy Crisis

Carolyn Kay Brancato

INTRODUCTION

If Alice in Wonderland were an energy economist, newly arrived in Washington, she might well be as disoriented as she was during her original trip. In a recessionary period, a Democratic President and a Democratic Congress, breaking with their Keynesian heritage, are busy balancing the budget by, among other things, shaving solar and conservation bank funds. Despite rampant inflation, the same group is deregulating natural gas and oil prices, while avoiding wage and price controls, last implemented by the Republicans under President Nixon. At least one large corporation, shelving any residual of its management's traditional adherence to laissez-faire, is practically on the government dole and everyone seems to want to eat from both sides of the magic mushroom to grow large (increase energy supplies) and small (reduce energy demand) at the same time.

It is wise, however, to recall that Alice, eating the mushroom, grows small than tall but ultimately remains in the same spot. Applying this metaphor to our energy policy, we might be compelled to ask the following questions:

(1) What is our country's energy policy?; and
(2) Is this energy policy internally consistent, or does one aspect counteract another?[1]

I. ENERGY INITIATIVES BEFORE PRESIDENT CARTER'S ADMINISTRATION

The range of current Congressional and Presidential initiatives in energy, taken together, might generally be construed as constituting the country's energy policy. To understand what this policy is and how it has been formulated, it is necessary to trace major legislative developments beginning with the period prior to the Arab oil embargo in October 1973. One often forgets that during the winter of 1972-1973, significant shortages of fuel oil developed in the Midwest and Northeast. Also, severe gasoline shortages materialized during the summer of 1973—months before the embargo. Thus, while OPEC pressure has been severe, significant market failures predate the first major confrontation with members of OPEC.

The Private Sector Has Traditionally Allocated Energy Supplies

President Nixon, however, was reluctant to use discretionary powers granted him by Congress to deal with these market disruptions[2]; this was most certainly in conformity with the implicit Federal energy policy, which had operated for decades:

> That policy has been to rely to the maximum degree on private enterprise to make the major investment, development, and pricing decisions affecting energy supply, to thus deliberately delegate to the private sector authority and responsibility for determining the evolving shape and direction of national energy policy as a whole.[3]

This laissez-faire policy rested on several assumptions:

(1) There are and will be sufficient domestic energy supplies.

(2) Energy supplies should be made available at the lowest possible costs and prices.

(3) Consumer choice among various kinds of fuels can be counted on to produce competition among supplies which in turn will provide incentive to keep energy prices low.

(4) Federal intervention should be limited to measures which would maintain a business and regulatory environment conducive to providing abundant and low cost energy. The federal role can also include providing a leading role in financing research and development of new energy technologies.[4].

Thus, prior to the early 1970s the private sector essentially shaped national patterns of energy production by making such fundamental decisions as: "the amount and direction of energy investments; the volume and rate of production; the direction and mode of distribution; the relative mix among various fuels; the prices at the wholesale and retail level; and the degree of national dependiture on imported fuels."[5]

Early Attempts Are Made to Establish a Federal Authority to Allocate Energy Supplies

Breaking with this tradition, in response to the market disruption of 1972-1973, Congress enacted legislation in April 1973, designed to enable the President to deal effectively with the shortages. The Economic Stabilization Act Amendments of 1973 (Public Law 93-28) provided the President with discretionary authority to allocate crude oil and petroleum products to "meet the essential needs of various sections of the nation and to prevent anticompetitive effects resulting from shortages of such products." The Act also authorized the President to establish priorities for the "use and for the systematic allocation" of petroleum products and crude oil to meet the objectives of the economic stabilization program then underway.

But President Nixon did not care to intervene in the market allocation mechanism, and his rather modest initiatives in establishing an energy department are further evidence of his reluctance to utilize discretionary authority thrust upon him by Congress. In April 1973, he established a Special Committee on Energy, composed of three of his closest advisors: John Ehrlichman, Henry Kissinger and George Schultz.[6] By June, amidst mounting discontent over fuel shortages,[7] the President set up the Energy Policy Office, which superseded the Special Committee. The Energy Policy Office, headed by the Governor of Colorado, John Love, served an advisory function only.[8]

When the Arab oil embargo came, the principal form of direct federal intervention in the oil market was the mandatory price controls on oil, initiated as part of the 1970 general economic stabilization wage and price control program. Congress, way ahead of the President, had already done considerable work on what was to become the Emergency Petroleum Allocation Act of 1973 (Public Law 93-159). The Act was passed by the Senate on June 5, 1973, was passed by the House on October 17th (on the day the embargo was announced), and was signed by the President on November 27, 1973. A landmark of potential direct federal intervention in the economic structure of the petroleum industry, the legislation made the voluntary allocation process mandatory "to insure distribution at equitable prices of petroleum products in short supply, both geographically within the country, to protect the market share of the independent sector."[9]

On December 4, 1973 President Nixon at last set up an energy office with more than an advisory function.[10] The Federal Energy Office, headed by William Simon, deputy secretary of the Treasury, had the responsibility of carrying out the allocation mandates of the Emergency Petroleum Allocation Act.

President Nixon Refuses to Accept Accelerated Federal Intrusion

But the President could only be pushed so far in accepting an expanded direct federal role in energy markets. The National Energy Emergency Act (S. 2589) introduced by Senator Jackson on October 18, 1973, the day after the Arab oil embargo, illustrates this point. After tumultuous debate, Congressional amendment, and Presidential counter-proposals, the bill was passed by the House and the Senate but was vetoed by the President on March 6, 1974.

Despite this failure, the legislation is important because during the course of debate, Congress and the President considered virtually the entire span of energy issues which subsequently have been dealt with by Presidents Ford and Carter and by the 94th, 95th and 96th Congresses. While the Emergency Petroleum Allocation Act provided for allocation of supplies at the production and distribution levels, this bill would have mandated allocation at the consumer level. The principal aim of the bill was to direct the President to draw up for the end-use consumer, a series of standby programs to impose mandatory rationing of petroleum products (principally of gasoline, fuel oil and propane), as well as mandatory conservation measures such as minimum home insulation standards, home and industry outdoor lighting and indoor temperature controls. Naturally, this further federal intrusion, even during emergencies, into the private sector was anathema to a President already reluctant to exercise his existing discretionary authority in the energy marketplace.[11]

Subsequent Energy Issues Are Defined

In addition to the rationing issue, the legislation, as amended, included proposals for: natural gas deregulation; prohibition of utility conversion from coal to oil; authorization for the President to order power plant conversion from oil to coal; exemptions from environmental regulations (stationary sources, Federal and State air and water quality standards) for certain energy projects; relaxation of auto emission standards, additional mass transit funding from the Highway Trust Fund, liberalization of nuclear power plant licensing; subsidized loans to homeowners and small businesses for those adversely affected by the legislation; tax deductions for energy conservation alterations, and creation of a Federal Energy Administration. Bur perhaps the most controversial aspect of the legislation was a provision added by the House to prevent price gouging. This involved use of Presidential Authority under the economic Stabilization Act of 1970 to specify prices for sales of crude oil, refined petroleum products, residual fuel oil and coal to prevent price gouging and to avoid windfall profits.

The National Energy Emergency Act Is Vetoed

President Nixon vetoed the National Energy Emergency Act on several grounds, some of which, in retrospect, seem surprising. Nixon objected to the unemployment benefits tied to the bill because of what he called the impossible task of determining whether the unemployment of each of the Nation's jobless workers is energy related. Furthermore, he argued that the subsidized interest loans would result in an outlay of government money not justified by anticipated energy savings. These objections are not surprising. But his principal target was the oil price rollback. Analysts of the Congressional Research Service have noted that the President's statements on this issue are an interesting revelation of the continuing ambivalence on the part of both Administration and Congress alike as to the present and future role of continuing imports.[12]

The first argument on this issue singled out as the basis for his veto was that it would discourage production—since domestic crude-oil prices would be set at such low levels that the oil industry would be unable to sustain its present production. His second argument against the rollback, however, was that it would *discourage imports*—since oil companies would be reluctant to import oil and gasoline that would have to be sold at prices far above the domestic prices. Finally, a rollback would take massive capital resources from the companies on whom we must depend for the investments in research and development in new energy sources needed to make us energy independent. Regarding this point, Nixon's veto message contained a surprising statement of support for a windfall profit tax measure with a plowback provision to exempt oil companies from the tax if they reinvest the proceeds in energy production.

> As we call upon industry to provide these supplies, I feel very strongly that we must also insure that oil companies do not benefit excessively from the energy problem. I continue to believe that the most effective remedy for unreasonably high profits is the windfall profits tax which I have proposed. That tax would eliminate unjust profits for the oil companies, but instead of reducing supplies, it would encourage expanded research, exploration and production of new energy resources. The Congress is holding hearings on this proposal, and I hope it will move rapidly towards passage. *I urge the Congress to enact this windfall profits tax as quickly as possible.* (Emphasis added).[13]

As we know, the Windfall Profit Tax, advocated by Nixon and others in 1974 and debated in each subsequent Congress, was finally enacted in 1980.

An Energy Independence Authority is Proposed

Despite President Nixon's reluctance to play a direct role in the energy market mechanism, President Ford, after replacing Nixon in August of 1974, placed considerable emphasis on economic incentives to increase domestic supplies of oil and gas and to stimulate production of alternative fuels to lessen dependence on oil and gas. The Energy Independence Authority, a creation of Vice-President Nelson Rockefeller, was proposed in October 1975. The proposed Authority was to dispense $100 billion in loans and other incentives for nuclear, synthetic and other fuels from domestic supplies. Significant Congressional support for this plan was voiced, especially from the Senate, although no legislation was passed.

Decontrol of Crude Oil Prices is Authorized

Also of significance during President Ford's tenure was enactment of the Energy Policy and Conservation Act (EPCA), Public Law 94-163, signed December 22, 1975. EPCA permitted, at the discretion of the administration, graduated decontrol of the price of "old" crude oil, which would rise to the world oil price as early as mid-1979. This represented a commitment to allow energy prices to rise to their replacement value, a policy at variance with the country's traditional commitment to produce energy at the lowest price. The legislation was to combine the conservation effect of higher prices with a production incentive effect to accelerate domestic production. Then, in August 1976, the Energy Conservation and Production Act (ECPA) was signed into law (Public Law 94-385). Together with a number of related bills enacted, these laws added to the already substantial base of Federal initiative[14] in the energy policy area by providing:

— authority for a strategic petroleum reserve;
— a decontrolled oil pricing policy;
— federal authority to force industrial and utility conversion from oil and gas to coal;
— an industrial energy conservation program (legislation enacted and funds authorized);
— automotive fuel economy standards.[14]

Finally, by the time the 94th Congress adjourned, the House and

Senate had reached agreement on substantial conservation and production tax credits, although legislation on these issues was not passed until the 95th Congress.

II. ENERGY LEGISLATION: 1977 TO THE PRESENT

Thus, when President Carter took office at the beginning of the 95th Congress, most of these cards relating to energy policy options appear to have been out on the table. And, much as in the croquet game Alice played with the Queen of Hearts, Alice might have observed in Washington some confusing shuffling of the players and the cards.[15] Indeed, Congressional Research Service analysis has found there were few elements in the goals, programs, and principles of President Carter's first energy plan, enunciated during his "moral equivalent of war" speech in April 1977, which could qualify as new. The Administration's quantitative goals were not very different from those proposed by President Nixon in his Project Independence and President Ford in subsequent energy messages. Most of the proposed programs had already been enacted or had received thorough Congressional consideration.[16] Furthermore, while President Carter's 1977 plan emphasized conservation, and his later plans, aired in April and July 1979, emphasized increasing domestic supplies, many of the critical elements of the game remain unchanged although the players have shifted.

President Carter Favors a More Direct Federal Presence

Perhaps the principal change the Carter Administration made with respect to energy policy was the President's articulation of the need for a direct Federal presence in energy matters. Whereas the preceding Republican administration advocated federal assistance in research and development by the private sector but rather grudgingly accepted even a limited role in the process to allocate energy supplies in emergencies, the Carter Administration staked out a more comprehensive Federal role such as that Congress had advocated during the prior administration. The five major pieces of legislation which comprise the National Energy Plan[17] are moving the country in the direction of mandatory rather than voluntary standards for a range of conservation techniques (appliance efficiency, building insulation, public utility rate reform, conversion from oil to coal for industrial and utility users). In addition, numerous tax incentives for solar and conservation methods provide direct tax rebate funding to individuals.

Carter's proposals during 1979 were more supply oriented, especially his proposed $88 billion program for synthetic fuels. While reminiscent of the Nixon-Ford approach to federal encouragement of large scale capital investment by the private sector, the Carter program contains more public-private partnership arrangements. Even in his advocacy of a windfall profit tax, Carter differed with Nixon as to how to channel these tax revenues into exploration and development. Nixon advocated tax forgiveness if the oil companies reinvested their excess profits, whereas in Carter's proposal, the federal government would collect the full amount of the tax revenues from the oil companies and then apply them to a range of government incentive programs.

Revenues from the Windfall Profit Tax are Directed at Increasing Supply as Well as Reducing Demand

Congress drew up an elaborate plan for dividing the projected $227.7 billion in windfall profits (Public Law 96-223, signed April 1980) between individual and business incentive programs to reduce demand and to stimulate supply. The following are the major provisions:[18]

* The existing tax credit for solar, geothermal, and wind energy property is increased to 40 percent of the first $10,000 of expenditures, (from 30 percent of the first $2,000 and 20 percent of the next $8,000) and equipment used to generate electricty from these renewable sources is made eligible for the credit.

* The Secretary of Energy is directed to evaluate six new items including passive solar equipment for possible eligibility for the residential energy tax credit, and specific standards are set for such evaluations.

* The extra business investment tax credit on solar, wind, and geothermal energy property is increased from 10 percent to 15 percent for 1980 through 1985, and solar process heat equipment is made eligible.

* The extra 10 percent investment credit for equipment to produce a solid fuel from biomass is extended from 1982 through 1985; and the extra 10 percent credit for equipment to convert biomass to alcohol for fuel use is extended to 1985 if the primary source of energy for the converting equipment is a substance other than oil, natural gas, or products of oil or natural gas.

* The existing exemption from the 4-cents-a-gallon excise tax on gasoline allowed blenders of alcohol with gasoline is extended from 1984 to 1992. Where the excise tax exemption does not ap-

ply, gasohol blenders are provided an income tax credit of 40 cents for each gallon of alcohol. Other breaks for gasohol were also agreed upon.

* Subject to various conditions, producers of certain alternative energy sources get a tax credit of $3 per barrel of oil equivalent, adjusted for inflation; the sources are oil from shale and tar sands, natural gas from certain nontraditional sources, synthetic fuels (other than alcohol) from coal, gas from biomass, steam from sold agricultural byproducts, and processed wood.

* Solid waste disposal facilities eligible for financing with tax-exempt industrial development bonds now include certain property used primarily to convert fuel derived from solid waste into steam as long as such property and that used for collection and processing of the waste is owned by the same person. Interest on an obligation used to finance a solid waste disposal facility and a related electric energy facility is also tax exempt under certain circumstances.

* Interest on industrial development bonds used to finance renewable energy property is now tax exempt in States that meet certain legal requirements.

* Small-scale hydroelectric facilities, including those of public utilities, are provided an extra 11 percent nonrefundable credit. The generating equipment must have an installed capacity of less than 125 megawatts and be installed at the site of an existing dam or at a site that does not involve the use of a dam or other water impounment structure.

* Under certain conditions, tax-exempt industrial development bonds may be used to finance hydroelectric facilities at existing dam sites or at sites where no dam or other water impoundment is involved.

* A new 10 percent energy credit through 1982 is provided for "co-generation" equipment added to an existing boiler or burner in which less than 20 percent of the annual fuel consumed is accounted for by oil or natural gas.

The Windfall Profit Tax Act also provides, among other things,[19] important financial relief from rising energy costs in the form of:

* Assistance to lower income families for heating and cooling costs is provided by means of an authorization of $3.115 billion for fiscal year 1981 (through block grants to States) with a general statement of intent that 25 percent of all windfall profit tax revenues will go to aiding lower income households.

Synfuels Legislation Is More Supply-Oriented

While the Windfall Profit Tax Act contains a mixture of proposals both to increase supply and reduce demand, the Synfuels legislation,[20] although it too contains both types of measures, is financially weighted much more heavily in favor of increased supply.

Title I of S. 932 deals with the development of synfuels under the Defense Production Act of 1950. Here the powers given to the executive to direct elements of the national economy to ensure that the defense establishment's needs for vital supplies such as energy are strongly advocated by the Congress and are embraced by the President, rather than rejected. S. 932 establishes an independent Federal entity called the United States Synthetic Fuels Corporation (SFC) with a national goal of producing 500,000 barrels of crude oil equivalent per day by 1987, increasing to 2 million barrels per day by 1992.

Basic provisions of Title I are:[21]

* The SCF Board of Directors is to be composed of 7 members appointed for seven-year terms by the President. In addition a six-member advisory committee is to be composed of the Secretaries of Energy, Interior, Treasury, Defense, the Administrator of EPA and the Chairman of the Energy Mobilization Board, assuming one is established.

* SFC's initial authorization is $20 billion (Carter wanted $88 billion, the House $3 billion and the Senate $20 billion). this $20 billion is subject to appropriations ($18.8 billion has already been appropriated in the FY 1980 Interior - Energy Departments Appropriations Act, Public Law 96-126). The financial resources available over the SFC's 12-year lifetime would be $88 billion, subject to appropriations by joint Congressional resolution under expedited procedures.

* The legislation empowers the SFC to provide financial assistance to the private sector for Commercial synfuel projects in the following order of decreasing priority:

1. Purchase agreements, price guarantees, and loan guarantees, up to 75 percent of the project costs;

2. Loans up to 49 percent of estimated project costs, although loans for up to 75 percent of the project costs could be extended if necessary to ensure the financial viability of the proposed project;

3. Minority equity interest, in a joint venture in which the Federal Government could provide up to 75 percent of project costs, in synfuels projects started prior to the approval of the comprehensive plan.

* Before the SFC could award loans and joint ventures, it must determine that purchase agreements, price guarantees, and loan guarantees either will not adequately support a viable project or will restrict available participants in the synfuel project. The SFC is prohibited from providing more than 15 percent of its obligational authority to any one synfuel project or to any one person or corporation.

* The conference agreement also authorizes the SFC to build up to three government-owned synfuel plants, subject to congressional veto. The SFC plants, which would be government-owned but contractor-constructed and -operated, would be authorized only prior to the approval of the comprehensive plan, only for one-of-a-kind facilities which use domestic resources, and only if no participant could be found who would be willing to proceed under one or more of the SFC financial incentives available to private investors. The SFC must divest itself of control of the plants within five years of acquisition.

* The agreement provides that any synfuels acquired by the SFC through purchase agreements, joint ventures, or government construction projects, must be offered first to the Defense Department for national defense needs before being offered to other federal agencies and the private sector.

The Synfuels legislation also provides for fiscal 1981 and 1982 a total of $1.2 billion for biomass and gasohol development activities. The goal is for gasohol to equal at least 10 percent of estimated U.S. gasoline consumption in 1990. Another $250 million is authorized for loans and loan guarantees for up to 75 percent of the capital cost of urban waste facilities to produce energy.

Renewable energy initiatives are to be undertaken as part of Title IV of S. 932, called The Omnibus Solar Commercialization Act of 1980.

This provision creates new incentives for the use of renewable energy resources and sets up a pilot program to promote local energy self-sufficiency. These provisions are:[22]

* The Energy Department is directed to coordinate all solar and conservation information distribution activities funded by DOE and report annually to Congress on the status of these activities.

* The Energy Department is directed to use a 7 percent discount rate and marginal (replacement) fuel costs in calculating the life-cycle costs of conservation and solar investments in federal buildings. This would have the effect of making the cost-benefit analysis of such investments more favorable.

* The Energy Department is directed to establish a three-year local energy self-sufficiency program, with a $10 million authorization for FY 1981. The program would demonstrate energy self-sufficiency through the use of renewable energy resources in one or more States.

* The National Energy Act is amended to expand the definition of small-scale hydropower facilities eligible for financial assistance from DOE from 15 megawatts to 30 megawatts. The Federal Regulatory Commission would be permitted to exempt very small hydropower facilities (less than 5 megawatts) from certain licensing requirements.

S. 932 Contains Solar and Conservation Bank Provisions as Well

S. 932, the Energy and Security Act of 1980, also creates a Solar Energy and Energy Conservation Bank within Department of Housing and Urban Development to provide subsidized loans to persons making energy conservation improvements or installing solar energy equipment. The following table shows the annual authorizations.

	Fiscal Year Authorization (millions of dollars)				
	1981	1982	1983	1984	TOTAL
Conservation:	200	625	800	895	2,530
Solar:	100	200	225	___	525

Source: Congressional Record, June 19, 1980, p. S 7405.

The conservation bank loans include maximum subsidy levels to be made based on a sliding scale of household income compared to the median income in the area. The scale is

Income Level of Borrower	Percentage	Maximum Subsidy Levels			
		1 Unit	2 Units	3 Units	4 Units
80% of area median	50	Up to $1,250	Up to $2,000	Up to $2,750	Up to $3,500
80 to 100% of area median	35	875	1,390	1,915	2,440
100 to 120% of area median	30	750	1,200	1,650	2,100
120 to 150% of area median	25	50	800	1,100	1,400

Source: Congressional Record, June 19, 1980, p. S 7407.

Solar loan subsidy ceilings are also established as follows:

Solar Loan Subsidy Ceilings

Household Income As a % of Area Median Income	Subsidy Rate	Maximum Subsidy per Type of Building		
		Single-Family	Two Units	Three to Four Units
80%	60% of cost up to:			
80% - 160%	50% of cost up to:	$5,000	$7,500	$10,000
Over 160%	40% of cost up to:			

Source: Democratic Study Group, Fact Sheet No. 96-40, Conference Report on Synthetic Fuels, June 20, 1980, p. 12.

Utility Conservation Programs under the National Energy Conservation Policy Act (NECPA) are revised to allow utility companies to finance directly residential conservation improvement as well as to extend the program to multifamily unit and small commercial establishments. Other provisions of S. 932 include:

* Weatherization funding is to be increased from $800 to a maximum of $1,600 per person;

* A total of $85 million in Federal loans and loan guarantees is authorized to promote exploration and confirmation of geothermal reservoirs;

* An interagency task force is established to study acid rain;

* The Federal government is required to commence filling the Strategic Petroleum Reserve at a minimum average rate of 100,000 barrels per day.

III. THE MAJOR ELEMENTS OF OUR ENERGY POLICY

With this historical perspective in mind, we can begin to answer Alice's first question: What is our country's energy policy? Several trends emerge:

(1) *Pre-embargo market disruptions and a continued OPEC threat have stimulated a desire to achieve self-sufficiency through increasing supplies and reducing demand.*

Table 1 and figure 1 illustrate our well-known petroleum production, consumption, and import history. Domestic petroleum consumption has increased from 14.7 million barrels a day (mmb/d) in 1970 to a peak of 18.85 mmb/d in 1978, and declined somewhat to 18.43 mmb/d in 1979. U.S. domestic production fell from 11.3 mmb/d in 1970 to a low of 9.74 mmb/d in 1976, and has increased to 10.17 mmb/d in 1979. Meanwhile, in 1979 we spent $56.73 billion on imported oil, compared to only $2.74 billion in 1970. Our imports have grown from 23.3 percent of our consumption in 1970 to 45.7 percent in 1979.

After the Arab oil embargo of 1973, the threat of increased prices became a major force motivating legislative action. Since the U.S. international oil companies were able to circumvent the embargo by buying elsewhere in the world oil pool, the threat of disruption as a major motivating force occurred later. Some analysts have linked it to the aftermath of the Iranian Revolution, although others have challenged the theory that gasoline shortages in 1979 were related to this event. Nevertheless, with the Soviet adventures in Yemen and Afghanistan, and the militant anti-American cast of the Khomeini regime in Iran, the perception has been heightened of potential disruptions of oil passing through the Strait of Hormuz. And, although we are less vulnerable than our allies to such a disruption (see Table 2), this threat, coupled with substantial OPEC price increases during the latter half of 1979, has galvanized support for a massive supply-oriented program to increase domestic supplies as well as some additional support for demand-reduction programs. But while consumption and import levels appear slightly reduced (Table 10-1), we are far from achieving self-sufficiency. Even if the synfuels program, which is the largest of our supply-oriented programs, were entirely successful, this would reduce our present level of imports by only 2 million barrels of oil equivalent by 1992, less than 25 percent of our current level.

TABLE 10-1

U.S. PETROLEUM STATISTICS

	Apparent Consumption[1] (million barrels a day)	Domestic Production[2] (million barrels a day)	Total Imports[3] (million barrels a day)	Import Value[4] (billion current dollars)	Imports As A Percentage Of Consumption[5]
1970	14.70	11.30	3.42	2.74	23.3
1971	15.21	11.16	3.93	3.35	25.9
1972	16.37	11.18	4.74	4.36	29.0
1973	17.31	10.95	6.26	7.74	36.2
1974	16.65	10.46	6.11	26.26	36.7
1975	16.32	10.01	6.06	25.06	37.2
1976	17.46	9.74	7.31	32.11	41.9
1977	18.43	9.86	8.81	42.01	47.8
1978	18.85	10.27	8.36	39.60	44.4
1979	18.43	10.17	8.41	56.73	45.7

1. Apparent consumption is derived by summing production, imports and withdrawals from primary stocks of refined products and subtracting exports. Essentially a disappearance measure, not "actual" consumption.

2. Domestic production is measured at the wellhead and includes lease condensate and natural gas plant liquids.

3. Total imports includes crude oil and refined petroleum product receipts into the 50 States and the District of Columbia. Included are imports for the Strategic Petroleum Reserve.

4. Import value includes crude oil and refined products.

5. Imports as a percentage of consumption is derived by dividing the volume of total imports by consumption.

Prepared by Jeffrey P. Brown, Congressional Research Service from the following sources:
    U.S. Department of the Interior, Bureau of Mines, Mineral Industry Surveys, Petroleum Statement Annual. U.S. Department of Energy, Energy Information Administration, Energy Data Reports, Petroleum Statement, Annual and Petroleum Statement, Monthly. U.S. Department of Commerce, Bureau of Census, U.S. Imports for Consumption and General Imports, FT 246.

U.S. PETROLEUM STATISTICS
1970 to 1979

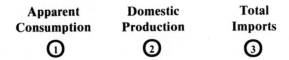

| Apparent Consumption ① | Domestic Production ② | Total Imports ③ |

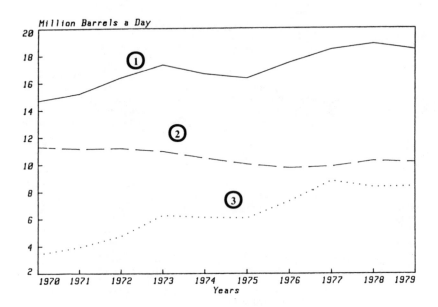

Prepared by the Congressional Research Service

TABLE 10-2

CRUDE OIL IMPORTS TRANSPORTED VIA THE STRAIT OF HORMUZ[1] AS A PERCENTAGE OF CONSUMPTION — 1973 and 1979

(thousand barrels a day)

	1971 Crude Oil Imports Via Hormuz	1973 Domestic Consumption	1973 Proportion of Consumption Via Hormuz	1979 Crude Oil Imports Via Hormuz	1979 Domestic Consumption Via Hormuz	1979 Proportion of Consumption
United States	620	17,308	3.6%	1,947[2]	18,488	10.5%
Canada	315	1,597	19.7%	258[2]	1,775	14.5%
France	1,533	2,219	69.1%	1.360	2,107	64.5%
Italy	1,327	1,515	87.6%	967[3]	1,607	60.2%
Japan	3,411	5,000	68.2%	3,240	5,173	62.6%
United Kingdom	1,575	1,958	80.4%	642	1,690	38.0%
West Germay	855	2,693	31.7%	803	2,664	30.1%

1. Oil-exporting nations that ship a significant portion of their petroleum exports through the Strait of Hormuz are Bahrain, Iran, Kuwait, Qatar, Saudi Arabia, and the United Arab Emirates.

2. January to November 1979.

3. January to September 1979.

Prepared by Jeffrey P. Brown, Congressional Research Service from the following sources: Energy Economics Research Limited. World Oil Trade. London, December 1979. U.S. Central Intelligence Agency. International Energy Statistical Review. Washington, April 23, 1980.

(2) *A direct Federal presence in energy markets has gradually become more acceptable.* Cooperation between the public and private sector is now common, especially where the government is the risk taker (for example, in the synfuel loan guarantee program), but there is still resistance to direct federal intrusion which would disturb patterns of end-use consumption.

While passage of the synfuels legislation marks the most substantial commitment yet to a cooperative venture to increase energy supply, there remains some deep-rooted ambivalence in the Congress and executive branch about the role of the Federal Government in dealing with energy problems. One area of ambivalence concerns the Federal role as regulator of end-use consumption. President Nixon vetoed legislation calling for standby rationing; President Carter has advocated it. It is clear, however, that the consumption reduction programs which have had the most widespread acceptance are not the rationing programs, but those with tax credits (Energy Tax Act of 1978 portion of National Energy Plan and Windfall Profit Tax Act of 1980) or subsidized loans for conservation and solar energy (synfuels legislation of 1980). This type of demand reduction allows maximum flexibility to the user and appears to be more ideologically acceptable to a society unfamiliar with significant end-use restrictions. Perhaps this factor has contributed to the great difficulty the Administration has had in fashioning a rationing program acceptable to Congress, as well as a program for building efficiency standards (BEPS). In addition, one might speculate that tax and subsidized loan programs may satisfy a better-defined constituency than programs which impose standards and have economic beneficiaries who may be reluctant to wait for a longer term payoff.

The ambivalence of a Federal presence in energy is also apparent from the demise of the proposed "fast track" Energy Mobilization Board (S. 1308). The proposed three member, full-time board and the President would be able, if created, to ask Congress to exempt a priority energy project[23] from laws blocking its construction. The Board would have the power to supersede Federal, State, and local laws.[24]

While President Carter strongly pushed this legislation, Congress added numerous steps to provide legislative safeguards. Despite these, the House voted down the bill on June 27, 1980. It is interesting that this time, unlike during the Nixon-Ford Administration, it is the President who is seeking more expansive, direct authority and the Congress which is checking that authority. The alliance to defeat the legislation might also have made Alice wonder. Conservatives largely concerned about usurping legitimate State and local responsibilities, but also concerned about the additional bureaucratic steps the Board might add, joined with liberals who feared expediting energy projects would damage the environment to the profit of major energy producing companies.

The legislation was defeated on two basic grounds, each related to an increased federal role. First, there was opposition to the concept that the federal government, even by joint action of the President and the Congress, should be allowed a discretion to override the amalgam of laws on the books. Second, there was a fear that the legislation would actually add steps to the review process and thereby would increase the federal presence in a different way.

(3) *It is now more acceptable to price energy at its replacement value instead of subsidizing its cost.* This pricing policy has the dual purpose of reducing demand, to the extent it is price elastic, and of increasing supply. Certain income distribution plans (for example, weatherization programs, windfall profit tax revenues for energy assistance to low income persons) have been passed to lessen the regressive effects of this revision in pricing policy. But the transition has not been smooth.

Recent Congressional opposition to the President's import fee illustrates that only a limited price increase will be tolerated, even in the pursuit of self-sufficiency. Pursuant to his authority under the Trade Expansion Act of 1962, as amended, the President can ask the Secretary of the Treasury to determine whether a commodity is entering the country "in such quantities or under such circumstances as to threaten or impair the national security." If such a finding is made, then the President may take such action, and for such time, as he deems necessary to adjust the imports of the commodity.

Effective March 15, 1980, President Carter imposed a fee of $4.62 a barrel on imported crude oil and $4.20 a barrel on imported gasoline. The fee was intended to increase only the price of gasoline and not the price of other petroleum products, such as home-heating oil. Reasons advanced to explain why Congress opposed the fee include: (1) the fee is, in reality, a 10-cent a gallon gasoline tax indirectly levied because the President lacked Congressional support for a direct tax; (2) Congress must protect its right to authorize taxes; (3) the mechanism for levying the fee is cumbersome since it involves imposing the fee on refiners and rebating according to the quantity of petroleum they import; (4) there is no way to limit the price increase to gasoline products and prevent a price increase on heating oil; (5) the fee will only save 2 percent of imported petroleum products, or 80,000 barrels a day, and this would be achieved at a cost of over $350 per barrel; and (6) the fee is a "blatant revenue-raising device" intended to bring in $10.7 billion to balance the budget.

Certainly issues of relative powers of the President and of the Congress to tax were important in the defeat of the import fee struck down by the court,[25] although Congress struck down the measure even before the court ruled.[26] In discerning Congressional motivation for this act, one can not overlook resistance to adding 10 cents a gallon to gasoline prices which had already risen considerably during the most recent year. An important issue raised during the Congressional debates was that the Department of Energy had not analyzed the comparative costs and benefits of the import fee, nor had it contrasted the fee to alternative and perhaps cheaper means of supplying or conserving an equivalent amount of oil.

Thus, while Congress has gone along with President Carter's oil decontrol plan which results in significant price increases on oil, there seem to be limits to congressional approval of price increases on petroleum products. Perhaps in the case of the oil import fee, the artificial taxing mechanism on top of the market-pricing structure was too much. Also while decontrol could be construed as a demand reduction technique, the import fee would have produced little demand restraint, but sizeable budgetary revenues at the expense of congressional taxing power. But the signal is clear: Congress will permit energy prices to rise, but not too much at any one time.

The debate over incremental natural gas pricing is another example of a certain ambivalence over policies to allow rising energy prices. Decontrol of natural gas prices was begun in accordance with the National Energy Act in 1978. But marginal rates targeted to one group of users, the industrial sector, are now under attack. Here the proposition that natural gas prices were too low and should be raised is not disputed, but there are growing pains associated with the distribution of such price increases, and objections raised are calling into question the whole concept of marginal cost pricing. These kinds of adjustments are understandable; one would not have anticipated an easy transition in energy pricing policy from the decades of low cost and subsidized rates to current high market rates, especially in an atmosphere of mistrust over the role of OPEC and of the international energy companies.

(4) *In terms of dollar values authorized to increase energy supply or to reduce energy demand, it appears that the supply side is ahead, especially considering the large synfuels subsidy.* In attacking the energy problem subsequent Congressional and Presidential initiatives have made incremental additions to both paths. Presidents Nixon and Ford were more committed to increasing supplies. President Carter's 1977 initiatives were more conservation oriented, but his 1979 initiatives were overwhelmingly supply oriented. Congress has had to push the Nixon and Ford administrations towards conservation and ultimately restrain the Carter Administration from spending $88 billion on synfuels and from taking the country down a "fast track".

(5) *Our energy policy appears to be based on a "try-a-little-of-everything" approach with incremental gains both in increasing supply and in reducing demand.* To the extent that both approaches will lessen our vulnerability to OPEC pricing and disruption threats, the components of our energy policy appear consistent. The policy is, however, arrived at by small changes in all directions; thus, there is some dispute over whether adequate consideration has been given to formulating this policy based on improving overall economic efficiency. Ideally, this would necessitate analyzing, if the data were available, which method (e.g. solar or synfuels) gives the greatest return for the federal dollar expended.

In addition, even if agreement could be reached on which path to take, for example, the solar, the synfuels or the nuclear path, there is uncertainty about how to use federal expenditures to reach the goal. Considerable discussion in Congress is devoted to ascertaining the relative contribution of loan guarantees, investment tax credits, accelerated depreciation, individual tax credits, purchase price subsidies, and the like in stimulating investment decisions which will reduce energy demand or increase energy supplies. Congress has found loan guarantees to be a most popular form of assistance. In addition, the 1980 Energy Security Act (S. 932) provides, for the first time, a sliding scale of assistance for solar and conservation improvements based on median income levels; this is another form of leveraging the federal subsidy. Also, current proposals for industrial conservation incentives include scaling the incentive to the differential between earned and target rates of return.

But the study of what motivates investment decisions is fraught with uncertainty, especially in a world of very rapid energy-price increases. While we have some experience using energy price elasticity of demand estimates to predict changes in behavior brought about by small shifts in price, recent large price increases in energy tend to negate this method to determine future behavior patterns. And, like Alice who is either too big to fit into the little door or is too small to reach the key on the table, we are uncertain of the proportions of federal incentives needed for us to fit through the self-sufficiency door. Meanwhile, pressures to balance the budget substantially increase the need for efficient decision making.

A critical issue for our future energy policy is the appropriate degree of cooperation between the public and private sectors. It also remains to be determined whether the market pricing of energy will, by itself, create sufficient incentives to reduce demand or make alternate energy supplies significantly more competitive by the year 2000. We cannot predict where things will settle, but there is one thing certain: like the White Rabbit, we are running late for an important date.

Notes

1. In other words, are we, like the Dodo and other birds Alice encounters on the seashore, merely running around in circles? See Carroll, Lewis, Alice's Adventures in Wonderland and Through The Looking-Glass, Collier Macmillan Publishers, London, Eleventh Printing 1977, pps. 41-48.

2. Historical materials drawn from the Congressional Research Service series written for the Senate Committee on Interior and Insular Affairs: Congress and the Nation's Environment, Environmental and Natural Resources Affairs of the 93rd Congress (April 1975) and of the 94th Congress (January 1977) as well as Energy Initiatives of the 95th Congress (May 1979).

3. Dave Gushee and Frances Gulick, "Energy and Fuels, Introduction," in Energy and Natural Resources Actions of the 94th Congress, op. cit., p. 9.

4. Ibid. pp. 9-10.

5. Ibid. p. 12.

6. Executive Order 11712.

7. Severe gasoline shortages were reported. Congress received testimony that over 2,000 retailers and distributors of gasoline products had been forced out of business by July 1973.

8. Executive Order 11726.

9. "Legislation Enacted," Environmental and Natural Resources Affairs of the 93rd Congress, *op. cit.,* p. 7.

10. Executive Order 11748.

11. The President, however, did seek more wide-reaching reaffirmation of delegated authority under the Defense Production Act to decide when an emergency existedas well as how to allocate essential commodities.

12. Environmental and Natural Resources Affairs of the 93d Congress, *op. cit.,* pp. 92-92.

13. *Idem.*

14. Needless to say, Congressional and Executive Branch follow-up for certain of these programs has been mixed. For example, purchases for the Strategic Petroleum Reserve have not followed a smooth course, and funds appropriated and expended for conservation have been less than those authorized.

15. This is not, as it might seem, a mixed metaphor. There *were* cards in the croquet game between Alice and the Queen of Hearts. The cards functioned as wickets and scrambled about to meet the hedgehog which was used as a ball. See Carroll, *op. cit.,* pp. 99-110.

16. Congressional Research Service, Overview Commentary on the President's National Energy Plan, prepared for The Committee on Interior and Insular Affairs and the Committee on Interstate and Foreign Commerce, U.S. House of Representatives, 95th Cong. 1st Sess., November 1977, p. 3.

17. The Natural Gas Policy Act (95-621), the Powerplant and Industrial Fuel Use Act (95-620), the National Energy Conservation Policy Act (95-619), the Energy Tax Act (95-618) and the Public Utility Regulatory Policies Act (95-618).

18. See Gelb, Bernard, The Oil Windfall Tax Act: A Summary of What it Means to Consumers and Small Businesses, Congressional Research Service Report, April 25, 1980.

19. Other tax provisions not related to energy matters include changes in estate taxes, the taxes paid on dividends and taxes paid on certain inventory evaluations.

20. S. 932, the Energy Security Act of 1980, passed by the Senate on June 19, 1980 and by the House on June 26, 1980. Signed by President Carter on July 4, 1980. "Synfuels" are synthetic fuels which have been converted from other materials such as coal and biomass. There are no commercial-sized synfuel plants now operating in the United States, although the SASOL plant in South Africa produces 2,000 barrels a day of crude oil equivalent.

21. See Congressional Record, June 19, 1980 beginning on S. 740 and Democratic Study Group, Fact Sheet No. 96-40, Conference Report on Synthetic Fuels Energy Security Act, June 20, 1980.

22. *Idem.*

23. "Energy Project" is defined to include all non-nuclear forms of projects for increasing energy supplies, including alternative fuels such as solar energy, increasing conservation, and for developing new technology.

24. If the committees with jurisdiction over the law to be waived agreed, then both Houses would have a chance to vote on the waiver. If both chambers agreed within 60 days and the President signed the measure, then the exception would become law. In a two-year Congressional session, only 12 projects would be eligible for waivers (Congressional Record, June 27, 1980, pp. H 5783-5784).

25. On May 13, 1980, U.S. District Court Judge Aubrey Robinson ruled the fee illegal and enjoined the administration from carrying out its program. The judge ruled that the program which constitutes a tax is an attempt to circumvent past congressional refusals to impose a gasoline tax.

26. Congress limited the President's authority under the Trade Expansion Act to prevent him from imposing fees on oil.

11 Toward a Secure Energy Future

Richard J. Stone

I. INTRODUCTION

The United States represents only five percent of the total world population. Yet it is the largest global producer, consumer and importer of energy—producing 22 percent, consuming 28 percent and importing 17 percent of all the energy produced in the world.[1] Energy now permeates almost every aspect of our daily lives from business, housing and transportation, to our leisure time and the quality of the environment in which we live. The cost of energy is the single largest contributor to the rate of inflation, the balance of payments and the value of the dollar on the international money market.

What we do—or fail to do—today about energy will have a profound and lasting impact not only on our own lives, but on the lives of our children and grandchildren for generations to come. Right now, Americans are paying a high price for past failures to act on energy. We are paying in the form of double-digit inflation, high unemployment, and day-to-day uncertainty about energy supplies and our very security as a nation. We can no longer afford the stop-gap approach to energy that characterized the policy decisions of the Nixon and Ford Administrations in the late 1960s and early 1970s. Their shortsighted responses to long-term questions have only increased the magnitude of the task before us.

We are making progress. Over the past three and one-half years, we have struggled toward a national consensus on energy policy. On November 8, 1978, President Carter signed the National Energy Act into law, over one and one-half years after he introduced this legislation in the Congress.[2] The Act represents our first comprehensive national program for energy. As President Carter said after passage of the Act, "We have declared to ourselves and to the world our intent to control our use of energy, and thereby to control our own destiny as a Nation.[3]

*The author is indebted to Casey Nikoloric for her invaluable assistance in the preparation of this article.

Since then, we have taken several bold new steps on energy. A number of these measures are contained in the Energy Security Act of 1980, recently signed by the President after more than a year of negotiation in the Congress.[4] Others include the decontrol of oil and natural gas and a number of major conservation initiatives.

These steps recognize that the United States does not exist in an energy vacuum. A long-term solution to our energy problem requires close cooperation with our Western European and Japanese Allies. The increasing interdependence of Allied economies, coupled with the threat that foreign oil dependence poses to the security of the Free World, mandates such a global approach. For this reason, President Carter has directed a strong United States initiative in the international energy arena.

This paper first will outline the nature of the energy problem we face, and then will describe the Carter program for addressing the problem. Finally, we will look at the international effort led by the United States, and the progress we have made toward increasing domestic supply and diminishing demand.

II. OUR ENERGY PROBLEM

To understand the full dimensions of our energy problem, we must begin with the realization that the world has come to the end of an era of "energy innocence." This has been a period in history characterized by the delusion that oil would remain forever inexpensive and readily available. Such a premise is irrelevant to the way we will live in the 1980s and beyond. It must not guide decisions about our energy future.

The Rise of Petroleum

Since the discovery of the first major American oil field in Pennsylvania in 1859, the availability and price of oil have dominated energy use here in the United States and abroad.[5] Early in the 1900s, oil began to replace coal as the major energy supplier to our economy. In 1900, the petroleum share of total U.S. energy consumption was only 2 percent, while coal supplied 71 percent of our energy needs. By 1930, the balance had shifted to 23.1 percent for petroleum and 58 percent for coal. In 1950, petroleum overtook coal and provided 38 percent of our total energy requirement, while coal supplied 36 percent. The petroleum share increased steadily through 1976, reaching 48.3 percent in 1977. Oil's share has only recently declined under President Carter to its present 42 percent. This same trend was followed elsewhere in the industrialized world.[6]

One of the most important reasons for this shift in consumption patterns was price. Until the Oil Embargo in 1973-74, the industrialized nations were able to rely on inexpensive petroleum to fuel their expanding economies. In fact, here in the United States the cost of petroleum actually fell from $1.19 per barrel for domestic crude at the well in 1900, to $.97 in 1935, rising only to $1.22 in 1945, $2.77 in 1955, and $2.86 in 1965.[7] By 1970, U.S. industry was deriving fully 75 percent of its energy requirements from oil and gas.[8] Entirely new markets and industries were created by the steady flow of cheap oil. We built our nation's highway system, automobile industry and suburban lifestyle on the premise that an abundant supply of inexpensive petroleum would always be immediately available.

Imports Exceed Exports

The year 1948 brought an important turning point for the United States when, for the first time, American oil imports exceeded exports and we became a permanent net importer of oil.[9] Our margin of dependence on foreign oil widened steadily throughout the latter part of the 1900s. In 1950, oil imports represent only 7.2 percent of U.S. petroleum consumption. From 1960 to 1976, the share increased from 19 to 43 percent.[10] The gap between domestic supply and demand grew at a particularly rapid rate during the 1970s, when domestic oil production peaked at 11.3 barrels per day in 1970 and President Nixon abandoned oil import quotas in 1973.[11] Since President Carter took office in 1976, the figure has declined to its present rate of 42.2. percent.[12]

The Oil Embargo

The Oil Embargo of 1973-74 demonstrated clearly that the United States was no longer in control of its own energy future. We learned the hard way that our dependence on foreign oil could be used as a political weapon against us. The United States was attacked on two fronts—supply and price. The embargo resulted in immediate nationwide petroleum shortages as the daily flow of nearly 500,000 barrels of oil (nearly 10 percent of total U.S. oil consumption) stopped arriving at American ports.[13] And for the first time in its history, the Organization of Petroleum Exporting Countries (OPEC) refused to negotiate prices. Instead, it set them. Saudi Arabia quadrupled the price of light crude following the Embargo—from $3.01 per barrel before the October 1973 Arab-Israeli War, to $11.65 per barrel on January 1, 1974.[14] During the Embargo, the United States experienced a $60 billion drop in the Gross National Product,[15] rapid acceleration in the Consumer Price Index (which rose 11 percent from 1973 to 1974);[16] and a drop of several billion dollars in the nation's world trade account balance.[17]

Instead of attacking the heart of the problem, however, the Nixon Administration looked only at its symptoms—choosing to treat the embargo as a temporary problem, an unanticipated and isolated event. Until the National Energy Act was introduced, no serious step was taken to move away from our dependence on foreign oil. And while our dependence increased, so did OPEC Prices, rising from $11.28 per barrel in January 1974 to $12.80 in 1977 and $18.64 in 1979. OPEC prices for April of this year were at $29.40,[18] and we can anticipate another $2 increase this summer as a result of the June OPEC meeting in Algiers.[19]

The High Price of Energy Dependence

This year it is anticipated that the United States will spend $90 billion to pay its imported oil bill, ten times what we spent on imports in 1973.[20] To bring this alarming figure into focus, consider: $90 billion is more than the total assets of the Ford Motor Company, General Motors and IBM, combined. $90 billion is more than all of the companies on the Fortune 500 list earned in 1979.[21] A $90 billion expenditure in one year is an outlay of $10 million per hour, $240 million per day and $400 per year for every man, woman and child in the United States.[22] At the average annual rate of increase in foreign oil expenditures, in just twelve years we will have exported a cumulative cash total equivalent to the present trading value of all the stocks listed on the New York Stock Exchange—over one trillion dollars.[23]

The impact of our alarming 42 percent dependence upon foreign oil is critical. However, the United States is considerably less dependent on foreign oil than the rest of the Free World. Western Europe is 63 percent dependent on foreign suppliers to meet its oil requirements, while Japan produces virtually none of the petroleum it uses.[24] The fact is that the rest of the world, like the United States, has not yet appreciably substituted other fuels for oil.

Several related issues add to the severity of our energy situation. First, our foreign sources of supply are exceptionally unstable. Of total Free World reserves—estimated at 545 billion barrels—fully two-thirds are located in Persian Gulf countries.[25] These nations provide 62 percent of all the oil in world trade today. Thirty-one percent of the foreign oil consumed in the United States is produced there.[26] In 1979, Saudi Arabia was our largest supplier at 1,346,000 barrels per day; Nigeria was second at 1,059,000 barrels per day; and Algeria and Libya were next at 1,272,000 barrels, combined.[27] Outside the Persian Gulf, only Mexico and the North Sea are expected to substantially increase current levels of production by 1985.[28] So it is clear that dependence on the Persian Gulf for oil will be a fact of life for the foreseeable future.

Northern Africa, the Middle East and, particularly, the Persian Gulf are areas of extreme political risk. In the last three decades, the Middle East has experienced six wars, twelve major revolutions and countless acts of terrorism, assasinations and territorial disputes.[29] This continuing instability already has impacted the world energy picture. In late 1978, the overthrow of the Shah curtailed production of more than 10 percent of the world's oil supply, and halted American purchases of Iranian oil.[30]

Second, the supply line through which Persian Gulf oil must pass is extraordinarily vulnerable to interruption. The Achilles Heel of today's world oil market is the tiny Strait of Hormuz, the narrow passageway to the Persian Gulf between Iran and the Arabian Sea through which one-half of all the oil sold to the Free World must pass in trade.[31]

Third, the major oil consuming nations face increasing and dangerous competition for less product. Except in Alaska, United States oil production has been declining steadily since 1971. Even production potential in OPEC is expected to peak in 1990.[32] Many Middle East nations have made recent decisions to cut back on production to ensure continued price escalation. At the recent OPEC meeting in Algiers, all OPEC nations except Saudia Arabia agreed in principle to cut total OPEC production by nearly 2 million barrels per day.[33]

Production cutback decisions also have resulted from the limited capacity of several Mid-East countries to absorb the growing revenues generated by the sale of their petroleum. Cutbacks in the economic development plans of some producing countries, a desire to conserve oil in the ground in anticipation of even higher prices, and a reduced inclination to accumulate assets elsewhere are additional factors in reduction decisions. And so, even payment of exorbitant oil prices cannot guarantee that the United States, or any other nation, will have a continued supply of foreign oil.[34]

Perhaps of greatest concern is the imminent entry of a new competitor. The Soviet Union, which has been a traditional exporter of oil, is expected to become a net importer in the next five years.[35] This fact has clear and present national security implications for the United States. While world attention has focused on the occupation of Afganistan, located less than 1000 miles from the Strait of Hormuz, other Soviet initiatives in the Middle East may be equally significant and indicative of anticipated need for new supplies of oil in the near future. These include the presence of a Russian fleet in the Indian Ocean and the Arabian Sea, the stockpiling of Soviet weapons in Libya and Syria, the massing of troops along Iranian borders, and Soviet presence in North and South Yemen and Ethiopia.[36]

Finally, we harbor a basic diseconomy in our patterns of energy use. The availability of cheap oil up until 1973, firmly has imbedded energy waste into every aspect of our society. More recently, inefficient patterns of energy use actually were encouraged by the artificially low oil prices maintained by controls under the Nixon and Ford Administrations. It is estimated that 30 percent or more of the energy consumed in the United States could be saved by more efficient use of available supplies.[37]

In short, America has become dangerously dependent on petroleum, a limited resource. Our major foreign supplies are concentrated in the hands of a few producing countries located in one of the most volatile areas of the world. It is an area upon which we exert little or no influence. Competition in today's tight oil market jeopardizes our traditionally close ties with the other consuming nations of the world. Our inefficient use of available energy further exacerbates an already difficult problem. This is the essence of our energy problem—and it is as critical a problem as we have faced in this century.

III. OUR ENERGY SOLUTION

Our mission for the 1980s is clear. We must reduce substantially our dependence on foreign oil and substitute a diversified, domestic energy base. We must reduce overall demand for oil through conservation. And finally, we must work with our friends abroad to implement a global energy strategy.

The Transition Away from Oil

With strong United States leadership, we anticipate that a worldwide transition in energy use patterns will occur in three phases.[38] For the next five years, from 1980 to 1985, the Free World will continue to rely heavily on oil. During this period, the most readily available source of additional energy will be conservation. Increases in the use of natural gas, coal and uranium will help reduce oil consumption in the near-term. Some non-OPEC countries, principally Mexico and the United Kingdom, are expected to increase oil and gas production. These increases will help maintain a stable energy supply and demand balance.

In the medium-term, from 1985 to 2000, the world will begin to make more significant moves away from oil. During this period, we will begin to more fully develop a number of options for reducing the demand for oil. These include coal and coal-derived synthetic fuels; solar and other renewable technologies; oil shale; unconventional gas supplies and nuclear power. In the medium-term, we also will see continued improvements in the efficiency of energy use.

Finally, beyond the year 2000, the world will move further toward renewable energy resources and nuclear technologies, consistent with health and safety requirements. Such technologies may include advanced solar storage and use systems as well as nuclear fusion. Our long-term objective is to develop renewable and essentially inexhaustible sources of energy.

The three phases of this transition away from oil have one feature in common. No single energy source, no one method for restraining demand and no single technological breakthrough will resolve the problem posed by our petroleum dependence. For the next several decades, we must explore all of our options. In the United States President Carter has undertaken a broad spectrum of initiatives to accomplish this important transition.

Conservation

In short-term, the most effective way to reduce our dependence on foreign oil is through aggressive efforts to maximize our energy efficiency. Conservation is the cleanest, least expensive and most readily available source of energy. In the past three and one-half years, we have begun a wide range of conservation efforts, including energy conservation standards, financial assistance and tax credits, promotion of ridesharing and driver efficiency, and mandatory conservation programs for federal facilities. More is being undertaken by industry as part of our voluntary programs.

More than three billion dollars for conservation activities by the Solar Energy and Energy Conservation Bank will be authorized over the next four years (excluding solar programs).[39] In addition to direct conservation payments to low-income tenants, energy conservation subsidies will be available through the Bank for residences and commercial buildings. Other programs include the establishment of energy efficiency standards for buildings and appliances, as well as the evaluation of similar standards for industrial equipment. Proposed Building and Energy Performance Standards (BEPS) are now under public review in the required comment period. Application of BEPS is expected to result in an average 35 percent reduction in energy use for residential and commercial structures.[40]

Additionally, under the Weatherization Assistance Program, some $200 million in housing weatherization assistance funding has been authorized for 1980.[41] To date, over 72,000 homes have been weatherized.[42] We also have begun an extensive program of energy audits and conservation measures for schools and hospitals throughout the country. Over the next five years, the program will pay for more than 58,000 energy audits, award more than 45,000 technical assistance grants and sponsor over 14,700 energy conservation improvements for institutional buildings.[43]

Because the transportation sector of the economy accounts for 53 percent of the petroleum consumed in the United States, the President is making a special effort in this area.[43] Personal motor vehicles in this country burn about 5 million barrels of oil a day.[44] On the average, we use more gasoline per capita than almost any other industrial nation. In 1979, our per capita consumption of gasoline was 11.7 barrels of gasoline, compared to 2.8 for France and 1.9 for Japan.[45].

The President has set a national goal to reduce gasoline consumption for 1980 by 5.5. percent.[46] Following a series of consultations with each governor, he has established voluntary consumption targets for every state. Together, these targets will achieve the national goal and reduce our gasoline consumption to 7 million gallons per day.[47] The targets can be made mandatory in the event of a severe supply interruption.

The President's transportation conservation initiative is designed to encourage ridesharing, vanpooling, driver efficiency and adherence to the 55 m.p.h. speed limit. Measures include sponsorship of Driver Efficiency Teach-Ins across the country, a program to achieve a 10 percent increase in federal vehicle fuel economy by the end of 1980 and the creation of the National Ridesharing Task Force to encourage private sector and public participation in ridesharing.[48]

Decontrol and the Windfall Profits Tax

President Carter took the first critical step toward a realistic national energy program early in 1977 when he introduced the Natural Gas Policy Act (part of the National Energy Act).[49] This Act brought to an end the dual pricing system of national and intra-state markets for natural gas and established in its place a uniform, nationwide market. Final passage of the Act over a year and a-half later put an end to the shortages of the previous winter and set the stage for the decontrol of crude oil to follow.[50]

A second series of Presidential initiatives was launched on June 1, 1979 to decontrol all domestically produced crude and abolished artificially low prices in existence since President Nixon imposed controls in September, 1973.[51] Decontrol began with the release of controls on the price of newly discovered oil (oil discovered since January 1, 1979). At the same time, 80 percent of all "marginal" oil (oil produced from wells with an output of 10-15 barrels per day) was permitted to be sold at a substantially higher, but still controlled price. In August, 1979, the President announced the immediate decontrol of "heavy" crude oil. As a result, by the end of this year domestic crude prices will reach $30 per barrel, compared to $13 per barrel in 1979.[52]

Controlled prices had discouraged conservation and inhibited exploration for domestic oil and development of other sources of energy. This led to the rapid escalation of the inflation rate because of our increased reliance on foreign oil. Recent estimates indicate that decontrol will result in reduction of foreign oil imports by 2 million barrels per day and an increase in domestic oil production of 1.2 million barrels per day by 1990.[53] Exploratory activities already have expanded to the highest levels experienced over the last 25 years, and the number of rotary drilling rigs in operation has increased from a monthly average of 1,656 in 1976 to 2,850 for the first five months of 1980.[54]

Along with these benefits, however, decontrol brings side effects that public policy must address. Higher prices from decontrol have meant substantial windfall profits for producers and a reduction in the real incomes of users—especially hard-hitting to the elderly and others on fixed and low incomes.[55] Recognizing these inequities and to help finance his program of energy diversification, President Carter proposed the Windfall Profits Tax to use the "windfall" earnings from decontrol to ease transition to higher oil prices.[56]

The Windfall Profits Tax will raise $227.7 billion through 1990 and will recover more than half of the additional net revenues producers will receive as the result of decontrol. In addition to assistance to low income families, the tax will fund conservation, energy research and development, the expansion and improvement of our mass transit systems and development of an American synthetic fuels industry by the Synthetic Fuels Corporation.

The tax also provides assistance to low-income households to help them meet the rising cost of energy. First, a $1.2 billion Special Energy Allowance has been set up to be distributed through the states on the basis of energy-related need. Second, an Energy Crisis Assistance program will transfer $400 million per year to the states to help low-income households meet energy emergencies, including repair of home-heating facilities and payment of extraordinary heating bills.[57]

Synthetic Fuels

The President has set a national goal for production of 1.75 million barrels per day in oil equivalent from synthetic fuels by 1990.[58] America has vast reserves of coal, oil shale, heavy oil and tar sands, unconventional gas and biomass resources that will permit us to achieve this critical goal.

More energy is available in United States deposits of coal than in petroleum, natural gas, oil shale and tar sands combined. Known American reserves are about one-half of the total world reserves. Over 400 billion tons are considered minable under current economic and technological conditions. These reserves are distributed throughout the United States. About 20 percent are located in the Appalachian Region, 30 percent in the Central United States and 50 percent in the West. Technologies to produce synthetics from coal include coal gasification, the process used to generate fuel gas, and direct and indirect liquefaction to produce methanol, boiler fuel and feedstocks used in the production of petroleum-like products. Depending on the process, synthetic liquid or gaseous fuels equivalent to about 1.5 to 3 barrels of fuel oil can be produced per ton of coal.[59]

Next to coal, oil shale is the nation's largest fossil fuel resource.[60] Oil shale deposits located in the Green River Formation stretching across Colorado, Wyoming and Utah are estimated to contain 600 to 700 billion barrels of recoverable oil. These deposits hold 25 gallons or more of shale oil per ton of rock. Recovery processes currently under development include surface mining, in-situ retorting, modified in-situ recovery and multimineral recovery.[61]

The bulk of the nation's heavy oil resources are concentrated in California and the Mid-Continent. Industry is currently producing this resource, believed to be 75 billion barrels of oil equivalent, at the rate of 250,000 barrels per day. Ninety-seven percent of all known tar sand resources, or an estimated 30 billion barrels in oil equivalent, are located in Utah. Both heavy oils and tar sands can be extracted from rock through the use of heat, solvents or hot water.[62]

Unconventional gas reserves in the United States are estimated at up to 1500 trillion cubic feet for those resources under current exploration. These include Eastern shales, Tight Gas Sands, and Coal Bed methane. All are formations where gas exists in reservoirs of low permeability. Recovery involves fracturing the reservoir rock, a technically complex and heretofore economically unattractive process.[63]

Biomass resources include plant life, agricultural and forestry waste, and municipal waste and industrial waste from organic material. The size of the potentially available land-based biomass resource in the United States has been estimated at 12 to 17 quadrillion Btu per year. Biomass, however, currently provides only two percent of the nation's energy supply, about 1.5 quads per year. Biomass resources can be collected and burned directly or converted through processes including fermentation and anaerobic digestion into liquid, gaseous or solid fuels; chemicals; and energy-intensive products. Fermentation technology is in place today, for example, to produce 80 million gallons per year of ethanol from grain and sugar crops.[64]

Research and development initiatives for biomass are complemented by the President's Alcohol Fuels Program. This program aims to quadruple current gasohol production by the end of 1980 and reach a capacity of 500 million gallons by 1990. Such a production level could displace almost ten percent of our demand for unleaded gasoline. Incentives have been provided to make gasohol competitive at the pump, including the permanent exemption of gasohol from the four-cent federal excise tax and a $145 billion, two-year federal credit program of loans and loan guarantees to stimulate private investments in ethanol production.[65]

The Synthetic Fuels Corporation, proposed by the President in the spring of 1977 and recently approved by the Congress as part of the Energy Security Act of 1980, will establish an American synfuels industry by acting as its investment banker. Over the next four years alone, the Corporation will obligate more than $20 billion in price guarantees, purchase commitments, loan guarantees and direct loans for commercial synfuel projects. An additional $68 billion may be made available for synfuels development through 1995.[66]

In advancing the development of synthetic fuels as an acceptable alternative to petroleum, President Carter has been particularly sensitive to the environmental issues associated with progress in this area. Air quality, reclamation, and water quality and availability all must be addressed in a satisfactory manner. In the final communique of the 1980 Economic Summit in Venice, the President and the leaders of the Free World's industrialized nations renewed their pledge to "fast-track" the development of synfuels in a fashion that would not do significant damage to the environment.[67]

Commitment to Coal

Coal is the most abundant resource in the world. The energy content of U.S. coal reserves alone exceeds that of all the crude oil reserves in the Middle East. At current rates of consumption, domestic coal would last two to three hundred years. Yet our coal industry is operating at only 20 percent of capacity. Coal accounts for more than 80 percent of domestic fossil fuels reserves, but supplies only 18 percent of the nation's energy needs. Excess production capacity is estimated at nearly 200 tons.[68]

Greater commitment to the use of coal in an environmentally acceptable manner is a key element in the world energy strategy for the future.

The International Energy Agency (IEA) has indicated, through its Coal Industry Advisory Board, that Free World production of coal could triple by the end of the century. The Board's "World Coal Study" concludes that, within a decade, expanded coal production would make a 25-percent cut in Western dependence on oil. This figure represents a cut of 6 million barrels per day in Free World imports from OPEC.[69] For this reason, together with the other major powers in the industrialized world, the United States has pledged itself to doubling the production of coal by 1990.[70]

In the spring of 1978, President Carter appointed the President's Commission on Coal to examine the potential for expanding the United States Coal Industry. The Commission issued its first major report in March of this year. The report concludes that coal can make a significant and cost-effective contribution toward reducing world dependence on foreign oil consistent with the requirements of the Clean Air Act.[71] On the basis of the Commission's work, President Carter has developed an aggressive four-part program to increase the production of coal. Full commitment to coal under the Carter program will give us a much needed source of domestic energy and generate some 59,000 direct jobs and 295,000 indirect jobs in mining communities.[72]

Part one of the President's coal program involves a ten-year, $10 billion effort to reduce utility oil and gas consumption by as much as one million barrels per day. This "Utility Oil Backout Program" will replace these fuels with coal and synthetics. Conversion to coal is already moving forward under the Power Plant and Industrial Fuel Act of 1978.[72] Under that Act, approximately twenty power plants have been ordered to cease burning oil, the first step in the conversion process. The burning of coal at these facilities alone will result in boosting coal demand by over 15 million tons per year at a savings of almost 175,000 barrels of oil per day.[73]

This year, the President has proposed new legislation, the Power Plant and Fuel Construction Act of 1980, which would mandate the conversion of an additonal 107 coal-capable plants at 50 generating stations by 1985. Conversion to coal at these stations could result in a 40 million ton increase in coal demand and the daily displacement of one million barrels of oil by 1990. The legislation would provide several billion dollars in federal grants to help utilities and their customers to pay for the high capital costs associated with conversion.[74]

Part two of the coal program involves rapid acceleration of coal leasing on federal lands. The antiquated policy of keeping federal lands off the market has forced the development of less economical tracts of land and has led to reduction in competition in the coal industry and increases in production costs.[75] After a careful examination of federal leasing policies, the Department of Energy recommended to the Department of Interior that sufficient federal acreage be leased in the next two to three years to ultimately produce 8 billion tons of coal, 50 percent more than is currently available on federal land. In response, the Department of Interior will this January hold its first federal coal lease auction since 1921.[76] The Department also has recommended that pending oil shale land exchanges be expedited on a priority basis, that the size of federal coal and oil shale tracts be increased and that firms be permitted to lease more than one shale tract per state.[77]

Part three of the coal program involves a heavier emphasis on coal exports. This commitment could result in the export of 80 million tons by the end of the decade and up to 380 million tons by the end of the century. The recently created National Coal Export Task Force will set targets for substantially increasing steam coal exports and make recommendations to the President for overcoming institutional barriers to growth of the coal export market. Even at the current price of $50 per ton, the export of 100 million tons of coal could earn this country $5 billion per year, and significantly help to reestablish a positive balance of trade.[77] The Task Force also is making a detailed examination of coal transportation issues, and exploring new ways to increase rail capacity and enlarge our already near capacity coastal port system.[78]

Finally, the Department of Energy has increased its investment in coal research and development to over $1 billion for fiscal year 1981, triple the amount spent on coal in 1978. A substantial portion of this funding will be devoted to the development of advanced pollution control techniques to support the President's coal expansion program. Clean, improved methods of burning coal are being developed through research and development projects which include work on a new process of mixing limestone with coal to remove impurities, and new ways to use coal in liquid form.[79]

Nuclear Energy

We must take care neither to overstate the potential of nuclear energy nor to underestimate its importance in the total energy mix. Last year nuclear energy met about 12 percent of our electrical generating needs. It is estimated that a 1000 megawatt power plant could displace 20,000 to 30,000 barrels of oil per day from an oil-fired utility system.[80]

The Carter Administration seeks to ensure that nuclear power remains a viable option to reduce our dependence on imported oil. The federal government's primary responsibility in this area is to ensure that nuclear power plants are built and operated safely. In April, 1979, following the accident at Three Mile Island, the President appointed the Kemeny Commission to investigate the incident and make recommendations on reactor safety. The Commission has completed its work and the President has adopted a number of its recommendations, including a proposed restructuring of the Nuclear Regulatory Commission and an increase in the number of federal inspectors at reactor sites.

Additionally, the President has taken steps to assure that nuclear waste is disposed of safely. A key element of the waste program is the establishment of a State Planning Council to consult with the states on the siting of repositories. Other initiatives include a thorough program to identify the nation's most suitable geological repository sites and the development of a National Radioactive Waste Plan, expected to be published in the near future.[81] These and other nuclear safety programs, together with the research and development program underway, will provide the technical knowledge from which a future generation of safe nuclear reactors may be developed.

Solar and Renewables

We also have moved to accelerate the pace of solar energy development. In addition to providing the conservation subsidies described above, the Solar Energy and Conservation Bank will provide assistance for owners and builders of residential and commercial structures who install solar equipment. Funding authorizations for the Bank's solar program will total $525 million over the next three years.[82] Recent estimates indicate that the Bank could finance over 100,000 retrofits during its first year. This program complements solar tax credits in the amount of $355 million already available through other federal agencies.[83]

Other ongoing renewable energy activities include expansion of wind energy systems, geothermal development, increases in hydroelectric capacity and continuation of the Ocean Thermal Energy Conversion (OTEC) program. The Department of Energy's Wind Energy Conversion program is already testing 200-kilowatt mechanisms in New Mexico, Puerto Rico, and Rhode Island. A fourth wind machine is under construction in Hawaii and the world's largest wind facility, a 2-megawatt machine in Boone, North Carolina, is also in place.[84]

The presence of large reservoirs of geothermal energy in the United States has already been established. These are thought to have the capacity to contribute 18.5 quads to our energy production by the year 2020. Programs are underway to stimulate commercial geothermal production from underground steam and water, hot saline fluids found in particularly porous rock formations and hot dry rock.[85] The OTEC program is designed to extract energy from waves, ocean currents and the salt levels of the ocean waters. Testing of various OTEC processes is underway in Hawaii, the Caribbean and off the coast of Florida.[86]

IV. INTERNATIONAL COOPERATION

The current global economic situation is characterized by a strong interdependence among the world's oil consuming countries. Increased demand for petroleum in one country can place enough pressure on world oil prices to affect inflation and growth in another. Conversely, lower petroleum consumption in one nation is a benefit which can be shared world-wide as price pressures are alleviated. This economic link makes a strong case for coordination of energy policy in the Free World. Under President Carter, the United States has taken the lead in developing a strong cooperative effort.

The International Import Ceilings

Since 1974, the cornerstone for cooperation among oil consuming nations has been the Emergency Allocation Agreement among members of the International Energy Agency (IEA), to which all major Free World Industrialized nations belong except France. This agreement provides a mechanism for sharing oil supplies in the event of a world supply shortfall of seven percent or more.[87].

In June, 1979, the leaders of these seven nations agreed at their annual Economic Summit in Tokyo to freeze 1985 imports at 1977-78 levels and to restrict oil imports in 1980.[88] At the December meeting of the ministers of the IEA, called at the request of President Carter, member nations also agreed on binding oil import ceilings for 1980 and 1985. The ceilings limited collective IEA demand for energy to 23.1 million barrels per day in 1980 and to 24.6 million barrels per day in 1985.[89] At the most recent IEA meeting in Paris, the ministers agreed to revise the 1985 ceiling downward to 22 million barrels per day.[90]

The 1980 Venice Summit

The 1980 Venice Summit confirmed United States leadership of the IEA energy effort. President Carter's program provided the framework for the joint energy policy adopted at the Summit. The overriding theme was energy, which dominated twelve of the 36 paragraphs of the final communique. The Summit leaders recognized the preeminent significance of energy on the world economy by stating that, "the economic issues that have dominated our thoughts are the price and supply of energy and the implications for inflation and the level of economic activity in our own countries and for the world as a whole. Unless we can deal with the problems of energy," they said, "we cannot cope with other problems."[91]

Special attention was focused on the progress of the U.S. energy program over the past three years. Clearly, anxiety about whether Americans would act on energy has turned to confidence that President Carter has constructed an energy program that is working and expanding. In fact, the Carter program now is recognized as a working model for the rest of the industrialized world. As a result, the seven industrial powers followed the U.S. lead by committing themselves to the development of the equivalent of 15 to 20 million barrels of oil per day from alternate energy sources. This amount represents approximately 50 to 60 percent of total OPEC production.[92] The final communique cited the world's need to "break the existing economic link between growth and consumption of oil"—already delinked here in the U.S.—by 1990.

The Summit endorsed the IEA estimate that member countries could cut collective imports to the 22 million barrel level for 1990 through a joint program emphasizing conservation and the development of coal, synthetics and, to a lesser degree, nuclear power.[93] To implement their program, member nations agreed that no new base-load, oil-fired generating plants be built and that conversion to coal and other fuels should be accelerated.[94] They also adopted measures to escalate residential and industrial conservation, the introduction of fuel-efficient cars into the market place and the development of alternate energy sources.[95] Quite evidently, the President's international initiative is paying off.

V. CONCLUSION

Over the last few years, we have moved faster and further toward establishing and implementing a comprehensive energy policy than at any other time in our history. There is more diversified energy activity in this country today than ever before. This activity is noticeable at every level of government and in the every day actions of the American people themselves. We have begun to rebuild the fabric of American society— refashioning the way we live around new energy realities. It is neither an easy nor a quick task, but we are making progress. We finally have recognized that we have to stop using foreign oil before we can stop buying it.

We have made substantial inroads toward reducing our demand for energy and, therefore, our demand for foreign oil. For the four weeks ending June 13, 1980, gross imports of crude oil and petroleum products together averaged 6.5 million barrels per day, 18.5 percent below the average for a comparable period last year. For the same four-week period, motor gasoline was supplied at an average rate of 6.6 million barrels per day, 8.7 percent below the average one year ago. At the same time, on June 13 crude oil stocks stood at 370.2 million barrels, 15.5 percent above the stocks level exactly one year earlier.[96]

On the supply side, total domestic energy production increased 4.8 percent from 1976 to 1979, up from 60 quads to 62.9 quads. Coal production is up from 685 million short tons in 1976 to 776 short tons in 1979.[97] New energy development activities have expanded as evidenced by the number of rigs in operation today and the high number of synfuel proposals pending with the Department of Energy. In response to a recent solicitation, industry filed over sixty proposals for feasibility studies or cooperative agreements for synfuel plants and nearly 100 proposals for converting solid waste into energy.[98] The Strategic Petroleum Reserve today is filled with 91.2 million barrels of crude oil, a six month's supply for use in the event of a shortage.[99]

And we have broken the link between economic growth and energy use. The United States energy coefficient, which is the ratio of growth rates for energy consumption and Gross National Product, dropped from 0.9 percent in 1976 to 0.024 percent in 1979.[100] Americans are moving beyond the energy problem to the opportunity inherent in our energy solution.

Throughout its history this Nation has risen successfully to meet new challenges. We are doing it again. We are energy rich and we are beginning to make the most of it. The time is none too soon, but it is better late than never.

Notes

1. Interview with Guy Carusso, Office of Market Analysis, Office of International Affairs, U.S. Dept. of Energy, July 1, 1980.
2. Public Laws 95-617, 95-619, 95-620 and 96-621.
3. Office of Public Affairs, U.S. Dept. of Energy, *A New Start: The National Energy Act.,* DOE/OPA-0042 at 1 (March, 1979).
4. Public Law 96-294.
5. For a detailed analysis of the world's petroleum industry, see M. Watkins, *The Emergence of a National Petroleum Industry,* Chapter Two (1980).
6. Statistics through 1973 are from: Resource and Land Investments Program, Council on Environmental Protection Agency, *Environmental Statistics 1978,* at 10. Beyond 1973, statistics are from: Energy Information Administration, U.S. Dept. of Energy, *Monthly Energy Review (May, 1980),* DOE/EIA 0035/05 at 6, (May, 1980).
7. Energy Information Administration, U.S. Dept. of Energy, *Annual Report to Congress,* Volume Two, DOE/EIA 00173 at 75, hereinafter *1979 EIA Annual Report.*
8. U.S. Dept. of Energy, *National Energy Plan II: A Report to Congress,* hereinafter NEP II, U.S. Gov't Printing Office Document No. 294-651/BID at 17 (May 1979)
9. *Energy Future,* at 18.
10. *Energy Future,* at 17 and 28.
11. *Monthly Energy Review,* (May, 1980) at 6 and 8.
12. *Energy Future,* at 17 and 27.
13. Energy Economics Research Ltd. and Middle East Economic Survey *International Crude Oil and Products Prices* (January, 1980) on file at the Office of Market Analysis, Office of International Affairs, U.S. Dept. of Energy.

14. *NEP II,* at 1.

15. *Economic Report of the President,* House Document No. 96–248, at 264 (January, 1980) hereinafter *Economic Report of the President, 1980.*

16. *Economic Report of the President, 1980,* at 323.

17. National Foreign Assessment Center, U.S. Central Intelligence Agency, *International Economic and Energy Statistical Review,* CIA Document No. ER IEER 80–002 at 10. (May 15, 1980)

18. New York Times, June 11, 1980, Section A at 1.

19. Address by Charles W. Duncan Jr., Secretary of the U.S. Dept. of Energy, at the National Press Club in Washington, D.C., June 26, 1980.

20. Calculations based upon figures published in *Fortune* at 53. (March, 1980)

21. Calculations based upon a total U.S. population figure of 220,584,000 *Economic Report of the President 1980* at 233.

22. Address by Charles W. Duncan Jr., Secretary of the U.S. Dept. of Energy, at the National Press Club in Washington D.C., June 26, 1980.

23. Testimony by John C. Sawhill, Deputy Secretary, U.S. Dept. of Energy, before the Senate Foreign Relations Committee in Washington, D.C., February 20, 1980.

24. Ibid.

25. Interview with Guy Carusso, Office of Market Analysis, Office of International Affairs, U.S. Dept. of Energy on June 28, 1980.

26. *1979 EIA Annual Report,* at 48.

27. U.S. Dept. of Energy, Secretary's Annual Report to Congress DOE/S–0010 at 1-4 (January, 1980) hereinafter referred to as *Secretary's Annual Report.*

28. *Energy Future,* at 31.

29. *Energy Future,* at 29-31.

30. St. Louis Post Dispatch, June 4, 1980, Section B at 1.

31. Testimony of John C. Sawhill, Deputy Secretary, U.S. Dept. of Energy, before the Senate Foreign Relations Committee in Washington, D.C., February 20, 1980.

32. New York Times, June 11, 1980, Section A at 1.

33. Testimony of John C. Sawhill, Deputy Secretary, U.S. Dept. of Energy, before the Senate Foreign Relations Committee in Washington, D.C., February 20, 1980.

34. *Secretary's Annual Report,* at 1-7.

35. St. Louis Post Dispatch, June 4, 1980, Section B at 1.

36. *Energy Future,* at 136.

37. See discussions in *Secretary's Annual Report,* Chapter One and NEP II, Chapter One.

38. Office of Public Affairs, Dept. of Energy, "What's in the New Synfuels Bill?" DOE Speakers Bulletin No. M–80–016 (June 24, 1980), hereinafter "What's in the New Synfuels Bill".

39. *Secretary's Annual Report,* at 2-3.

40. *Secretary's Annual Report,* at 2-9.

41. Office of Public Affairs, U.S. Dept. of Energy, "Additional Accomplishments," DOE Speakers Bulletin No. M–80–019 (June 30, 1980).

42. Ibid.

43. The White House Press Office, "Fact Sheet on the President's Import Reduction Program," U.S. Gov't Printing Office No. 0–297–618, at 9 (July 16, 1979), hereinafter "The White House Fact Sheet".

44. "The White House Fact Sheet", at 9.

45. Testimony of John C. Sawhill, Deputy Secretary, U.S. Dept. of Energy, before the Senate Foreign Relations Committee in Washington, D.C., February 20, 1980.

46. "The White House Fact Sheet", at 1.

47. "The White House Fact Sheet", at 5.

48. "The White House Fact Sheet", at 1.

49. Public Law 96–223.

50. *Economic Report of the President, 1980,* at 109.

51. Public Law 93–621.

52. Office of Public Affairs, U.S. Dept. of Energy, "Speakers Bulletin," DOE Speakers Bulletin No. M–80–0167, (June 24, 1980).

53. *Economic Report of the President, 1980,* at 110.

54. Address by Charles W. Duncan Jr., Secretary of the U.S. Dept. of Energy, at the National Press Club in Washington, D.C., June 26, 1980.

55. *Economic Report of the President, 1980,* at 109.

56. Public Law 95–223.

57. White House Press Office, "Windfall Profits Tax: A Reality," (April 11, 1980), on file at the White House.

58. "What's in the New Synfuels Bill", at 1.

59. U.S. Dept. of Energy, *Synfuels: Status and Next Steps,* at B-1 and B-2, (February 1, 1980), (unpublished draft on file in the Office of the Assistant Secretary for Resource Applications), hereinafter *Synfuels: Status.*

60. *Synfuels: Status,* at A-1.

61. *Synfuels, Status,* at A-3,4.

62. *Synfuels, Status,* at D-1.

63. *Synfuels, Status,* at G-1.

64. *Synfuels, Status,* at E-1.

65. Interview with Bernie Greenglan, Director, Office of Alcohol Fuels, U.S. Dept. of Energy, July 3, 1980.

66. "What's in the New Synfuels Bill", at 3.

67. Text of the final communique of the Seven-National Economic Summit held on June 22 and 23, 1980, hereinafter "Venice Communique". Participants included the United States, West Germany, Japan, France, Britain, Canada and Italy. Text made available through the Associated Press (on file in the Office of International Affairs, U.S. Dept. of Energy).

68. Address by John C. Sawhill, Deputy Secretary, U.S. Dept. of Energy, before the American Mining Congress on May 5, 1980.

69. New York Times, June 25, 1980, Section D at 2.

70. "Venice Communique", at paragraph one.

71. The President's Commission on Coal, *Recommendations and Summary Findings,* at 3, (March, 1980).

72. Address by John C. Sawhill, Deputy Secretary, U.S. Dept. of Energy, before the American Mining Congress on May 5, 1980.

73. Public Law 95–620.

74. Address by John C. Sawhill, Deputy Secretary, U.S. Dept. of Energy, before the American Mining Congress on May 5, 1980.

75. Senate Bill 2470, House Resolutions 6930 and 6947.

76. Interview with Dr. Bruce Lawton, Director of the Office of Leasing, Office of the Under Secretary, U.S. Dept. of Energy, July 3, 1980.

77. Address by John C. Sawhill, Deputy Secretary, U.S. Dept. of Energy, before the American Mining Congress on May 5, 1980.

78. Newsweek, June 30, 1980 at 46.

79. *Secretary's Annual Report to Congress, 1980,* at 4-3.

80. Address by John C. Sawhill, Deputy Secretary of the U.S. Dept. of Energy, before the United States Chamber of Commerce in Washington, D.C., on April 28, 1980.

81. The Report of the Kemeny Commission is on file at the White House Press Office.

82. "What's in the New Synfuels Bill", at 1.

83. *Secretary's Annual Report to Congress, 1980,* at 3-3.

84. *Secretary's Annual Report to Congress, 1980,* at 3-11.

85. *Secretary's Annual Report to Congress, 1980,* at 3-17.

86. *Secretary's Annual Report to Congress, 1980,* at 3-13.

87. *Economic Report of the President, 1980,* at 166.

88. Testimony of John C. Sawhill, Deputy Secretary, U.S. Dept. of Energy, before the Senate Foreign Relations Committee in Washington, D.C., February 20, 1980.

89. *Economic Report of the President, 1980,* at 167.

90. The Washington Post, June 24, 1980, Section A at 1.

91. "Venice Communique", at paragraph 1.

92. *Monthly Energy Review (May, 1980),* at 88.

93. "Venice Communique", at paragraph 1.

94. "Venice Communique", at paragraph 1.

95. Ibid.

96. Office of Public Affairs, U.S. Dept. of Energy, "Major Accomplishments of the Nation in Energy", DOE Speakers Bulletin No. M–80–015 (June 23, 1980).

97. *Monthly Energy Review (May, 1980),* at 4.

98. Address by Charles W. Duncan Jr., Secretary of the U.S. Dept. of Energy, at the National Press Club in Washington, D.C., June 26, 1980.

99. "Energy Accomplishments" (July, 1980), on file at the Office of Intergovernmental Affairs.

100. Office of Public Affairs, U.S. Dept. of Energy, "Major Accomplishments of the Nation in Energy" DOE Speakers Bulletin No. M–80–015 (June 23, 1980).

12 Responses to Oil Supply Vulnerability

Walter S. Baer

Oil Import Dependence and Vulnerability

The United States' dependency on imported oil poses two inter-related but distinct problems: 1) chronic economic losses and 2) vulnerability to supply disruptions. Each problem has its own characteristic time frame and requires a different set of policy responses.

We hear most about the chronic economic problems associated with import dependence. Our steady-state purchases of six to seven million barrels of oil and refined products daily from foreign suppliers results in large economic losses to the U.S. economy—losses from transfers of wealth to the producing countries, higher inflation, balance of payments problems, and the decline of the dollar. These problems are with us to-day and will remain with us for some time to come. They are also receiving high priority policy attention. The large synthetic fuels program recently enacted, conservation measures, and phased decontrol of oil and gas prices all represent attempts to do something about the steady-state dependence problem.

The second, closely related problem is our vulnerability to oil supply disruptions. It arises not because we import so much oil, but because so much of the oil we import comes from the politically unstable Middle East. Unlike the chronic economic losses from import dependence, oil supply disruptions are, of course, unpredictable. But a significant disruption will have very severe economic and political impacts on our allies as well as on ourselves. And, regrettably, while we have concentrated on synfuels and other measures to reduce import dependence, we have largely neglected responses to the vulnerability problem itself.

Reducing oil imports certainly helps, but it doesn't eliminate our vulnerability problem. What is needed is more in the nature of contingency planning to prepare us to deal with possible supply disruptions.

Dependence on Persian Gulf Oil

Figure 12-1 illustrates the international nature of the dependence and vulnerability problems. The United States imports about 3 million barrels per day (mbpd) from the Arab Oil Producing Countries (OPEC), but this represents only 17 percent of our total oil use. Our European allies and Japan are a good deal more dependent. France imports 83 percent of its oil from the Middle East, Japan 60 percent, and the industrialized countries as a whole—represented by the Organization for Economic Cooperation and Development (OECD)—receive a little more than one-third of their oil from Arab producers.

This dependence will persist throughout the 1980s despite the measures that we and other countries are adopting to reduce oil imports. Rising prices will decrease demand and encourage new sources of supply, such as synfuels, but significant production of alternative liquid fuels is at least ten years away. The increased exploration for and production of oil and gas outside the Persian Gulf region will likely be matched by increased demand from developing nations. And despite price increases, we will be fortunate if U.S. domestic oil production does not decline significantly from its current level during the decade.

Likelihood of Future Oil Supply Disruptions

With our continuing dependence on oil from the Middle East, we and our allies face the very real prospect that these supplies could be cut off, rapidly and without warning, by any number of events in that unstable part of the world. For five years after 1973 we tended to view political uses of the "oil weapon," like the Arab embargo after the Yom Kippur War, as the primary threat. But the sharp cutback of Iranian oil exports in 1978-79 made us realize that supply disruptions can occur because of civil unrest largely unrelated to world markets or politics. Looking ahead in the 1980s, we see a host of unpleasant possibilities: civil unrest in other producing states, including Saudi Arabia; expansion of ever-present regional hostilities between Iraq and its neighbors; a new Arab-Israeli war; intervention of the Soviet Union; or even actions that are not under the control of governments, such as terrorist destruction of pipelines and port facilities, or a tanker accident that could block the Strait of Hormuz.

DEPENDENCE ON MIDDLE EAST OIL

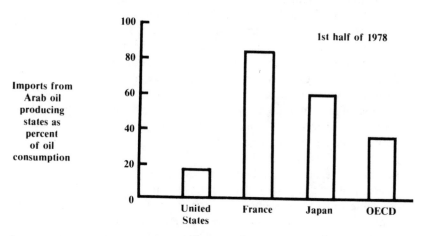

Source: International energy statistical review, CIA

Figure 12-1

Any one of these events may be relatively improbable, but taken together, they add up to a significant probability that something will go wrong. Over the next five years, it seems more likely than not that we will have to deal with a supply disruption at least as great as that caused by the shutdown of Iranian exports in 1978-79. And we should remember that those shortages, although never more than 5 percent of oil production, led to a doubling of world oil prices, long gasoline lines in the United States, and near panic reactions on the part of some consuming countries.

Economic and Political Effects of Disruptions

A significant oil supply disruption will bring both severe economic and political effects. As an example of a major disruption, consider the impacts if Middle East supplies were cut by 9 million barrels per day for a period of one year. This represents the approximate total of current exports from Saudi Arabia. Under current sharing arrangements worked out by the International Energy Agency, the United States would see its imports reduced by about 35 percent of the total cut, or by more than 3 mbpd.

Several groups inside and outside government have estimated the economic effects of such a disruption for the United States and other industrialized countries.* Under current conditions, for example, calculations by Henry Rowen and colleagues at Stanford University indicate

*These results must be considered speculative, because the economic models used to derive them were developed for incremental changes in inputs, not for the large dislocations associated with a disruption.

industrialized countries.* Under current conditions, for example, calculations by Henry Rowen and colleagues at Stanford University indicate that the United States would suffer a loss of some $200 billion, or seven percent of GNP, from such a disruption. Estimates made by the Congressional Budget Office are of a similar magnitude. The impact in Western Europe and Japan would be even greater: nine and ten percent losses in GNP, respectively, according to the Stanford estimates. The disruption would force immediate, dramatic cuts in industrial production and employment. Oil prices would rapidly rise to above $100 a barrel. The general economic dislocations would make last summer's gasoline lines seem like a pleasant diversion.

The political effects of such a disruption are less amenable to analysis, but even more sobering. A serious disruption would create internal stress within every democratic country whose outcome is unforseeable but not pleasant to contemplate. Internationally, we would certainly see great strains on an already shaky Western alliance. Consuming countries might rush to bid in the spot market for petroleum, as we saw in 1978, driving spot market prices through the roof. Individual consuming nations would also likely try to make separate, bilateral deals with those producers unaffected by the disruption. Depending on the political nature of the interruption, there could be demands for some kind of Western military response. The United States is now working to build greater capability to project forces into the region during a crisis. But we are obviosly not the only actor. The threat of direct Soviet intervention remains a major factor in any scenario. In the event of true chaos in the Persian Gulf, the Soviet Union might be tempted to intervene in order to "stabilize" the situation and restore oil supplies to Western Europe and Japan.

Preparations to Reduce Vulnerability

How, then, can we better prepare ourselves for dealing with future oil supply disruptions? Some measures to increase supplies and reduce demand in an emergency are listed below:

Preparatory measures for emergency supplies
- National petroleum reserves
- Private stockpiles of oil and refined products
- Standby production and distribution capacity
- Plans for emergency fuel switching

Preparatory measures for emergency demand restraint
- Standby gasoline rationing
- Emergency taxes or tariffs

These are largely standby measures or contingency plans, which again should be distinguished from longer-term policies to reduce dependence such as synfuels or conservation programs. They are also measures that, to be effective, should be closely coordinated among the industrialized, oil consuming nations. Although this paper principally addresses U.S. policies, vulnerability to oil disruptions is an international problem that requires international responses.*

A Strategic Petroleum Reserve (SPR)

The most important supply side measure is to build up a national oil reserve that can be rapidly pumped out for use during a supply disruption or other emergency. Such national reserves are helpful in several ways. They act as deterrents against a political embargo that might be threatened, for example, should war break out again between the Arab states and Israel. If a supply interruption does occur, petroleum reserves can help keep oil prices from exploding. The Stanford group estimates that with total OECD reserves of 2.8 billion barrels, the world price of oil would remain some $50 a barrel below its price without such reserves.

Knowledge that stockpiles are available should also lessen panic reactions by consumers, industrial firms, and governments themselves. This last point is of particular importance. If a disruption occurs, and no petroleum reserves are readily available, the United States and other governments may feel great pressures to take immediate political or military actions. The availability of reserves will buy time to plan more measured responses.

The International Energy Agency has recommended that each member nation store at least 90 days' supply of petroleum. For the United States, this represents approximately 750 million barrels (Fig. 12-2). Storage facilities for about 250 million barrels have been built, with a second phase underway to construct storage for another 280 million barrels. President Carter in the 1977 National Energy Plan called for an expanded, billion-barrel U.S. Strategic Petroleum Reserve (SPR). Recent studies conclude that even larger petroleum reserves would have positive cost benefit ratios. The SPR optimal size depends on one's estimates of the probability and extent of future disruptions. For the U.S., a ten percent probability per year of a large disruption (3 million barrels per day for one year) is enough to justify a SPR larger than a billion barrels. But at present, only 93 million barrels are actually in place, or less than a two week supply of imports.

---

*Discussion of current and proposed arrangements to share oil supplies in an emergency, through the International Energy Agency or some other means, is beyond the scope of this paper.

THE U.S. STRATEGIC PETROLEUM RESERVE (SPR)

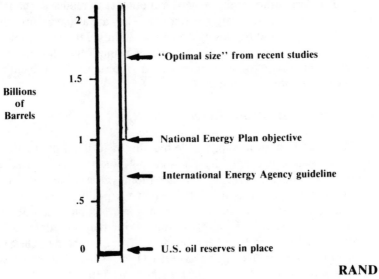

Figure 12-2

Expanding the Strategic Petroleum Reserve thus becomes an urgent matter for reducing U.S. vulnerability. The problems that have prevented SPR purchases since the Iranian crisis of 1978 are no longer compelling. Softness in the current world oil market would permit purchases without significant upward pressure on oil prices. Consequently, we should be in a better position than before to deal with Saudi Arabian objections to our filling the SPR.* In the recently passed Energy Security Act, Congress has mandated a resumption of oil purchases for the SPR, although at an initial level of only 100,000 barrels per day. This is far too low a rate to achieve a credible stockpile. At 100,000 barrels per day, it would take nearly twenty years to reach a 750 million barrel SPR. At least doubling that rate seems a sensible near-term objective.

---

*The impact of oil purchases on the federal budget has been another objection. Filling the SPR at 300,000 barrels per day (the rate established before the Iranian crisis) would add more than $4 billion to the annual budget deficit at current world prices. Off-budget financial arrangements probably can be justified for the SPR, however, especially since rising world oil prices suggest the likelihood of inventory gains from holding reserves. An SPR Administration separate from the Department of Energy could be given independent financial authority. Its securities would be backed not only by U.S. Government guarantees, but by barrels of oil in the ground, which could appeal to many investors. Alternative, a COMSAT-like, public-private corporation could finance and manage the SPR. This is much like what has been done in West Germany.

Plans to Use the SPR

A Strategic Petroleum Reserve has little value unless there are plans to use it in an emergency. In the past, some have considered the SPR for its deterrence or insurance value alone. According to this view, withdrawals from the SPR should be made only as a last resort, like collecting from an insurance policy only when the insured is dead or severely disabled. However, recent analyses conclude that the SPR's value is increased if it is used during small as well as large interruptions. Drawdown will both reduce panic among oil consumers and ease the necessary economic adjustments.

Regrettably, the United States has no established plan for SPR drawdown during emergencies. Establishing such a plan is thus a priority item for energy policy. The drawdown plan must include the criteria that would trigger SPR withdrawal, the withdrawal rate, and the methods for determining prices and allocations. Standby authority for implementing the plan should be established by Congress after thorough analysis, hearings, and public debate. The SPR drawdown should then proceed rather automatically during an emergency. This will enable the President and other top government officials to turn attention to the international security aspects of an oil emergency, rather than domestic fuel allocations.

Private Stockpiles of Oil and Refined Products

Policy should encourage building oil reserves in private hands as well as in the government-controlled SPR. Today, private stocks of oil and refined products are at an all-time high, due to inventory buildups in response to last year's shortages and price increases. Firms are estimated to hold more than 250 million barrels above minimum inventory levels, or about three times the amount of oil in the SPR.

Rising oil prices have given firms incentives to stockpile, but they see substantial disincentives as well. Government reallocation in an emergency is their principal concern. It's the old story of the ant who plans ahead versus the grasshopper who doesn't; but the ants fear that the grasshoppers will have more votes in Congress to reallocate supplies during a crisis.

Reducing the fear of government reallocation would be the most effective way to maintain or increase private stockpiles, but this is easier to preach than to put into practice. Regulations or legislation enacted now to encourage ant-like behavior could still be rescinded later by the grasshoppers. However, some specific initiatives can be undertaken. Holding costs for private inventories could be reduced by allowing private storage in the SPR—at least until it is filled with government purchased oil. Tax credits or other subsidies could be offered for private stockpiles. Subsidy of private stockpiles, of course, weakens the argument against their reallocation in the event of emergency.

Finally, the Energy Policy and Conservation Act of 1975 authorizes the Secretary of Energy to require importers and refiners to store up to three percent of the oil imported or refined in the previous year. Such mandatory requirements could be increased if other incentives fail.

Emergency Oil and Gas Production

There are limited, but still important opportunities to increase oil and gas supplies in an emergency. With world prices so high, little shut-in capacity exists in the United States (or anywhere outside the Persian Gulf region). A few fields in Texas that are currently non-producing because of state production ceilings could be tapped, perhaps contributing as much as 100,000 additional barrels per day. Other producing wells could be pushed beyond their optimal long-term production rates for a short period of time, if it were in the national interest to do so at the expense of long-term recovery.

More Alaskan oil could be shipped to other parts of the country by increasing the operational capacity of the Trans-Alaska Pipeline. The pipeline was designed to carry 2 million barrels per day but is currently working at less than 1.5 mbpd. Additional pumping stations would be needed to handle the increased flow. Even if the pumping stations are not justified by current economics, they could be subsidized and put in place on a standby basis as part of an emergency preparedness program.

Canada and Mexico could also produce more oil and gas during a disruption of Persian Gulf supplies. Their capabilities to expand exports rapidly are now limited by existing distribution as well as production facilities. However, Mexican President Lopez Portillo has indicated that Mexico should build excess capacity of about ten percent of oil production, or about 250 thousand barrels per day.* Greater increases in emergency export capacity by either Canada or Mexico would require multilateral agreements, and probably some front-end financing by consuming countries. For Mexico in particular, multilateral arrangements, perhaps through the International Energy Agency, appear more acceptable than an attempt to secure emergency supplies for the United States alone.

---

*David Ronfeldt, Richard Nehring and Arturo Gandara, *Mexico's Petroleum and U.S. Policy: Implications for the 1980s,* The Rand Corporation, R-2510-DOE, June 1980.

Emergency Fuel Switching for Electricity Generation

Contingency plans should be developed for generating electricity from other fuels during an oil supply disruption. The United States currently uses about 1.5 million barrels per day of oil for electricity generation, principally in the Northwest, Florida, Texas and California. With pre-planning, much of that could be replaced in an emergency.

If natural gas were available, it would be the replacement fuel of choice. Operating coal and nuclear power plants at higher-than-usual capacity factors would also help, if the extra power can displace that generated from oil-fired plants. This requires long-distance power transfers (known as "wheeling") between the coal and nuclear plants and the oil consuming utilities. Although most utilities now wheel power routinely, some may have to increase the capacity of their ties with other utility systems to prepare for emergency power exchanges. Changes may also be needed in some state utility commission rules and contracts among utilities to facilitate emergency wheeling.

Emergency Measures to Restrain Demand

A severe oil supply disruption will require strong measures to cut demand quickly. The general alternatives are direct government allocation of supplies, such as gasoline rationing, and price increases through emergency tariffs or taxes.

Any rationing scheme would be expensive and an administrative monstrosity. The recent Congressional hearings on the gasoline rationing plan developed by the Department of Energy clearly identify the problems. Gasoline rationing would take six months to a year to implent under the best of circumstances. The first round of coupon distribution would probably miss 15 to 20 million vehicles. The program would face millions of exemption requests, as well as continuing problems of counterfeiting and fraud. And the rationing bureaucracy created to deal with these problems would be difficult to disband once the emergency was over.

For these reasons, analysts generally favor emergency tariffs or taxes to cut demand and reduce the enormous additional transfers of wealth that would otherwise go to oil producers during a supply disruption. In principle, a tariff or tax would impose lower administrative and social costs than rationing, and presumably would be easier to remove after the emergency ended. A tariff, as opposed to a gasoline tax, has the advantage of applying neutrally to all oil products, but a tax at the refinery input level could give essentially the same results.

The size of the emergency tariff or tax would depend, of course, on the severity of the disruption. For the previous example of a total 9 million barrel per day cut (translating to a 3 mbpd loss for the United States under current sharing arrangements), analysts estimate that roughly doubling the present price would clear the market. This means a tariff or tax of $30-$40 a barrel.

The real issue is fairness. The chilly reception given recent proposals for a gasoline tax or tariff one-tenth that size shows how politically difficult it is to impose energy taxes on consumers, even as emergency standby measures. The windfall profits tax could serve to drain crisis-caused profits from domestic producers. Any new tariff or tax proposal to avoid windfall transfers to foreign producers must include means to rebate money directly to consumers. And the rebate mechanism must be quick, highly visible, and perceived as fair by the Amercian public.

Many rebate schemes can be suggested, including direct refunds to individuals via the federal income tax, payments to all households, or specific rebates to those affected most harshly. Whatever the rebate mechanism, Congress does not now seem ready to pass new energy taxes or tariffs, even on a standby basis. Still, it seems important to address these issues publicly, including holding Congressional hearings, to set the stage for more rapid response if it becomes necessary.

*   *   *

In summary, our vulnerability to oil supply disruptions is at least as serious as the more chronic economic problems of import dependence. Measures to deal with the vulnerability problem include:

- Increasing oil purchases for the Strategic Petroleum Reserve;
- Providing incentives for private stockpiling; and
- Developing emergency preparedness plans for

  — SPR drawdown
  — Electricity generation from gas, coal, nuclear
  — Emergency tariffs or taxes
  — Standby gasoline rationing.

This does not mean that we should delay current measures to reduce overall import dependence, such as the decontrol of oil and gas prices. But we must focus more clearly on the vulnerability problem itself as the most critical short-term energy issue for the United States and our allies. Only if we are prepared to weather oil supply disruptions in the next few years will we be able to devise longer-term solutions to our energy problems.

13 The Energy Crisis and the American Lifestyle

L.J. Cherene

"Lifestyle" is best defined as the technology of living—the activities, methods, and resources employed in the regimen of daily life. As such, it evolves from an accumulation of personal experience, cultural traditions, legal and institutional constraints, and, yes, even prices and scarcities of resources. The issues examined here are the implications of what shall be called the "American Lifestyle" to energy consumption and the implications of a coming age of expensive energy upon the future evolution of this life style.

The defining of the "American Lifestyle" (hereafter referred to as ALS) is frought with conceptual pitfalls and sociological controversy. For the purpose of this presentation, I shall define it as residing in a single-family, detached dwelling with a two-car garage surrounded by land. This residence is a base of household operations—a place to entertain guests, cook meals, and bed down for daily sleep. Most resources for these operations are obtained from outside the "home" by exchanging labor services for money at a place of employment, located in a region dedicated to commercial or industrial use, and exchanging money for resources at a commercial center. This is not to say that all Americans live this way, nor does it imply that no one else does. In fact, this life style, some times called "The American Dream," seems to have come to the United States from England, where the idyllic image of a cottage in the country far from the city with its rude activities of work and commerce is still strong to this day.

In contrast to the ALS is the Urban Life Style (ULS), characterized by neighborhoods of multiple-household, multiple-story residences, the ground floor usually dedicated to commercial use. The same exchanges of labor for money for goods takes place, but the locations of employment and shopping are intimately intermingled among the residences in such a fashion as to be close to what a practitioner of the ALS would call his "back yard". While the subsistence activities of eating and sleeping are carried out in these domiciles, entertainment usually takes place in a local cafe, cantina, or beer garden. Even reading is done not on a livingroom couch but in a local library. The term "urban" does not include high rise office buildings and large, self-contained commercial centers (so called "shopping malls"). Such specialization of land use is part of the ALS and denoted with the uniquely American phrase, "downtown". Neither does the "urban" rule out rural settlements; Italian and German agricultural workers, for instance, typically live in multiple family dwellings in a self-sufficient village and commute to the fields, leaving family behind to practice their roles in the ULS. This contrasts to the ALS where the farmer and his family lives in his single-family detached dwelling surrounded by his land and commutes to the general store, the gas station, and the Sears catalogue outlet that characterize the rural town.

The energy use implied by the ALS becomes evident when compared to that of the ULS. First, consider heating the home. In the U.S., this accounts for roughly 15% of all energy consumption. The typical dwelling of the ALS has six sides exposed to the elements and may be insulated with R19 rated material in the attic space. In contrast, the typical ULS residence has two and at most four sides so exposed, the remaining sides insulated from climatic variations by other living spaces—excellent insulation indeed. Further, the average size of the ALS dwelling is larger than that of the ULS, requiring proportionally more heat. The per capita impact on energy use is significant. Even summer cooling is less energy intensive in the ULS by virtue of the mutual insulation of living units.

But energy costs, even at today's prices, have a minimal effect upon the cost of the ALS when the institutional constraints are considered. In the U.S., interest payments are deductible from taxable income, rental payments are not, resulting in a government subsidy of the ALS. FHA, VA loans, and other such government programs are institutional monuments to the acceptance of the ALS as *the* way to live. Bankers, too, think, the same way; single-family dwellings are financed up to 90%, condominiums to 80%, and non-owner occupied apartments to 70%, and the interest rates reflect this same bias. The ULS is often disqualified from financial support via "red lining", North Boston being a case in point. In fact, many practitioners of the ALS view the ULS as "slum" or "ghetto" living, in reference to its high population densities. But these high densities are the source of economies, particularly with respect to energy. "Ghetto" refers not to population density but to homogeneity of the district—Black Ghettos and Jewish Ghettos exist side by side with the financial ghettos and single-family, detached dwelling ghettos of the ALS. In fact the ALS's tendency to produce homogenous ghettos devoted to housing, commerce, finance, and their subuses is the source of another energy diseconomy-transportation.

The image of the ALS city is one of gleaming towers of offices and shops entwined by the graceful curves of wide, sweeping expressways. The relationship between these elements is most complementary, in fact, skyscrapers imply freeway. Any geographical allocation to specialized uses demands a system for transporting people from one region to the other. It is no accident that the ALS creates commercial centers that appear to be high rise monoliths in a sea of traffic arteries and parking spaces. Here life only exists between 8:00 am and 5:00 pm. the huge facilities necessary to handle peaks of transportation at these times appear as vacant and forbidding wastelands for the rest of the day. When the source of personnel must commute from the low density housing of the ALS, the most cost effective mode of transportation is the automobile.

The automobile is a hallmark of the ALS for two reasons. Economically, its economies stem from the wide geographical separations between places of work, shopping, and residing. From the aesthetic standpoint, it is the extension of the single-family, detached dwelling—the single-family, detached transport module. Despite its efficiencies within the ALS, it is still responsible for some 15% of energy consumption in the U.S. Other forms of "mass" transportation between the specialized ghettos of the ALS consume a considerable share of energy as well.

The heterogeneity of ULS neighborhoods make such large allocations of energy for transport services unnecessary. Under these conditions, two feet are generally all that is necessary for most transportation needs, especially for shopping activities. High density generates enough demand per acre for most services to be located within walking distance of each domicile, dispensing with the need for either automobile or mass transit services. The role of the street is not a wide stream bed for treacherous flash floods of commuter traffic, but a narrow strip of open space where neighbors meet and interact while traveling from shop to shop. The street of the ULS is, in effect, the commons—a public foyer for each dwelling it serves.

Such heterogeneity of uses is foreign to the ALS. Modern city planning requires simplification for the sake of conceptual manageability, and specialized use is the first simplification made. Great swaths of land are ordained to be devoted to commercial, low density, middle density, and high density residential uses. This very act significantly increases the transportation needs of the community. Decentralization is the next step in simplified planning. Each building is designed in a vacuum, so the intricate relationship between uses and location that arise in the city of the

ULS are never considered in the city of the ALS. Here parking lots and low-profile, single-use buildings abound having no cornice line in common, humbled by great naked curtains of steel, glass, and stone, separated by vast "open spaces" of grass, parking lots and the accouterments of drive-in commercialism. Housing developments, too, are single use ghettos, empty during the daylight hours when adults work and children are schooled. Such developments are not mass produced packages released by cost minimizing monopolies to a captive market, but producers' reactions to what is in popular demand. Specialization has even created geographic rifts between demographic classes; the elderly, once the daytime caretakers of the community and located as close to the family as the back bedroom, now live in "retirement communities" that are a good twenty-minute drive from kith and kin. How many Btu's will a visit be worth in the future?

The keywords to the ALS are low density residences and homogeneity of land use. "We have everything you can possibly want within a thirty-minute drive" is the common boast of this life style's practitioners. But the cheap energy that made it all possible is no longer available. If today's accommodation is tomorrow's tradition, the trend toward heterogeneous ("mixed use") planning and high density living via the condominium may shift Americans away from the ALS to something more closely resembling the ULS. The transition may not be so hard as one might suspect. After all, very few people in the world today live in single family dwellings. Throughout the world a man's house literally is his castle; the masses live in apartments. In Paris, a good apartment is a family largess—a homestead handed down through generations. Not only would these people have it no other way, but they can afford it still.

14 Observations Concerning an Optimal Solution of the Energy Crisis

Dave Button

At the outset, I must note that nothing which follows is new, and literally millions of businesses and consumers act out the roles underlying the views expressed. The discussion is arranged as follows:

First are comments on some of the nomenclature commonly used today in discussions concerning energy. The point to be made is that many of the questions surrounding energy are begged from the start.

Next, the question which I think should be addressed will be answered and some implications will be drawn.

Then, an assessment of where the answers would lead us in terms of macro-economic impact, inflation, employment, productivity, consumers, etc. will be summarized.

Finally, the role and responsibility of the firm in the "energy industry" will be presented.

The best place to start is with the theme of this book, "Solutions to the Energy Problem: Alternative Approaches to the U.S. Energy Crisis," the question "Is there a solution to the Energy Problem?" In addition to these phrases—Energy Problem, Energy Crisis, and Energy Solutions, I would add Energy Security, Supply Vulnerability, and, of course, National Energy Policy should be added.

All these phrases lead to the notion that there is an unusual and undesirable situation out there which can and ought to be corrected. Furthermore, the liberal use of terms like national interest and national policy suggests that there is a premier role for the Federal Government to play in this supposed dilemma. Carrying this reasoning farther leads to the familiar claims of market failure and the need for government intervention, and finally, when the current prescription for government policy and action fails, a new improved federal remedy is sought.

If, instead, energy matters were viewed as common place market phenomena, the concerns would be different. If energy, in general, and petroleum, in particular, were viewed as rather ordinary commodities, then market adjustments would be expected and accepted just like the phenomenal adjustments which have routinely taken place in sugar, gold, and silver markets, just to name a few. This, of course, has not been the perception, but I suggest that if it had been in 1973 when prices changed and supply was interrupted, we would not have the government caused situation we face today.

I trust that I need not enumerate all of the regulations and programs which have been distorting domestic energy markets since 1971. It should be commonly accepted fact that the petroleum industry has had to bear a relatively greater injection of inefficient government regulation and taxation over a ten-year period than has any other major industry in the history of this country. An alternative now, and always, has been to let the market adjust as efficiently as possible; in other words, without government intervention. But for some reason, reliance on independent individual choice, the workings of a market, has been rejected.

While it is certainly true that the energy issue is exceedingly complex and multidimensional, it is inaccurate to state that it is not well understood. In particular, economists should understand the issue well and find its complexity and multidimensional nature far from unique. In fact, as far as the United States is concerned, petroleum could represent a reasonable real world example of a competitive price takers market. Granted the time frame involved is long, but the dynamics from the point of view of fundamental micro-economics principles are straight-forward.

Furthermore, economists should understand well it is the straight-forward competitive nature of energy and petroleum markets, rather than their complexity, which makes formation of government policies fraught with difficulty. After all, aren't the markets and series of transactions associated with virtually everything we buy and sell infinitely complex when compared to the models some of us try to build? And isn't it always difficult to form policies which improve something that is naturally efficient? It seems as though too many economists have forgotten why the principles governing efficient markets are taught in elementary courses. It is not because they are simple-minded, but because they are basic and powerful foundations for the rest of economics. When basic principles are violated, the search for explanations beyond that which is fundamentally obvious is of questionable value.

One More Caveat

There is no such thing as an "industry view." Firms may have views, but industries don't. For the petroleum industry, in particular, the views of individual firms are quite diverse. Getty's views derive from an appreciation of markets and what they can do to create opportunity and allocate resources efficiently, if left alone. Other firms within the industry take quite a different approach, advocating all sorts of government regulatory schemes. So those thoughts are, at best, in line with one firm's general philosophy, and there is no reason to pretend that firm has the same interests as its competitors. In fact, the reverse seems appropriate. It is important to recognize that the oil industry is not one giant monolith with a single mind. That type of characterization is far from true, and nothing could make it more clear than to force each firm in the industry to compete in the market place rather than in Washington, D.C.

The question that I wish to address is: "What shoud be done about energy?"

The answer to this depends on where the question is directed.

For consumers, the answer is purchase what tastes, prices, and income dictate.

For energy companies or oil companies, the answer is produce as much as possible, given price and cost constraints.

For the federal government, the answer is—do nothing, or, given the facts of life as they exist today—the federal government should deregulate every aspect of the energy market as soon as possible. I am well aware that removing the stain of over-regulation is no simple task. Nevertheless, economic logic and the record of failure clearly dictate that the appropriate direction is away from government involvement.

As for more specific national implications of no government market involvement, they are straight-forward, but the effects may need some elaboration.

Concerning prices, a no-government involvement policy would imply a more rapid decontrol of oil and petroleum product prices; in fact, immediate decontrol is both possible and desirable. This would have the effect of stimulating domestic production and promoting more efficient consumption choices. However, the full economic benefits would not be realized, due to the windfall profits tax, and so that legislation ought to be repealed. To advocate otherwise requires heroic assumption that the federal government can spend the tax funds more efficiently than the industry which generates those funds.

Also natural gas prices should be decontrolled immediately, particularly for all new contracts. Again, this would generate additional supplies while encouraging efficient use. It must be noted that the Natural Gas Policy Act of 1978, which is sometimes described as a gas price deregulation bill, falls far short of the mark and includes such typical government dictated absurdities as allowing gas produced from wells deeper than 15,000 feet to receive any price the market will bear while gas from 14,999-foot wells is regulated at less than half the market price.

The, government involvement position would also involve immediate removal of all allocation schemes such as entitlements and the supplier/purchaser freeze. Again immediate removal of federal programs is both possible and desirable. Their removal would allow crude oil resources to flow freely to those refiners who can produce products for consumers at the lowest real cost possible.

Application of a no-government involvement philosophy is easy when talking about prices and free exchange of private property. But, what about government programs which are designed to protect national security, like import quotas and stockpiling programs such as the Strategic Petroleum Reserve?

The import quota issue seems straightforward—none are needed. When applied to crude oil or specific products, quotas would unnecessarily raise the price of those commodities and resources to American consumers and business. It is interesting to note that the most often touted reason for a crude oil import quota system is to protect us from the disruptive effects of a supply cut-off. Yet, it seems to me that quotas represent a self-imposed cut-off and assure that many of the disruptive events which are feared, but are only speculative, will actually occur.

Another government program designed to protect against supply interruptions is the Strategic Petroleum Reserve. This type of government sponsored program has a lot of appeal, even to some strong advocates of private enterprise. Many feel that a free market would fail to produce an appropriate hedge against interruption. But closer examination will show no market failure and that the only thing that prevents an appropriate reserve from being stored is the federal government. It is the government's price controls, its obsession with false notions of windfalls and fairness, and its confiscatory programs of mandatory allocations which have prevented crude oil storage from occuring naturally. In a free market with no government involvement, the possibility of supply interruptions would represent a business opportunity. Insurance minded entrepreneurs would see the need to provide a hedge and would recognize that stored supplies would be highly valued in the event of an "unforeseen" cut-off. But given the past behavior of our government, who would be foolish enough to think that a private concern would be allowed to sell and keep the proceeds from an investment of this type. As a practical matter, the federal government has so distorted expectations, that even if it were removed from the energy arena, it would take quite some time to restore credibility. Only then would businessmen make the efficient decisions that would have occurred had the government never been involved in the first place.

There are, of course, government programs which give hand outs to corporations. A good example is the Energy Security Fund which is designed to subsidize synfuels. If synfuels have a place, and probably they do, normal market forces, if there is no government intervention, will bring them forth in a timely manner. The fact that we don't have large scale synfuel projects now relates to their cost relative to less expensive alternatives like foreign oil and the uncertainty created in part by the federal government's inclination to grant subsidies to certain pet projects. Can anyone be blamed for waiting to see which way the government favor will lean? And remember, its hard to justify a subsidy for something you have already done.

The list of awards is interesting. In the DOE's most recent $200 million synfuel give away, Union Oil gets $4 million to do an oil shale feasibility study. Given Union Oil is one of the largest owners of oil shale property, would not one think that it is in their interest to figure out what to do with it? Does Union Oil need a $4 million grant? (Of course it may be argued that the $4 million comes from Union's share of the windfall profits tax, but that just points out what's wrong with having the tax.) The Azusa Land Reclamation Company gets $1.9 million to build an electric power plant associated with methane recovery from land fills where the methane is produced using what the project's manager calls ". . . known technology; every item comes off the shelf". As a matter of fact, Getty Oil Company, without this subsidy, has been producing methane from land fills for five years. Without feasibility grants, etc., Getty was the one who pioneered the technology and methods which Azusa will be using to qualify for their award.

At this point, there is an obligation to point out the impact of a zero, or at least reduced, government role in the energy markets. Here the comments offered are general. In at least one oil company, the business is finding and producing oil, so the models developed and the areas where detailed attention is focused are relatively narrow. Detailed views of every aspect of the U.S. economy aren't developed. One of the nice things about markets is that, when guided by wealth maximization, it is not necessary for every participant to know much about the larger picture. Anyway, there are more studies and econometric model simulations out there than anyone can use, so one more view of the universe will probably not be missed by many.

The impact of the policy suggested here can be summarized quickly. Because economic waste and inefficiency promoted by government intervention would be reduced, the following directions, tendencies, etc. would develop.

1. Energy produce prices to consumers over the long run would be as low as possible.
2. Productivity in general would be increased.
3. Employment would tend to increase (except for those who work for the DOE).
4. OPEC prices would tend to become softer.
5. Inflation would be reduced due to increases in productivity assuming that money supply is unchanged. (Though oil price

changes have an immediate effect on the CPI statement, they have nothing to do with the inflation we are experiencing. Long-term, persistent inflation at the levels we have been seeing is caused entirely by the irresponsible monetary and fiscal actions of the Federal Reserve and Federal Government.

Finally, a brief picture of what one firm is doing in the energy market, such as it is, appears in order.

The firm's responsibility is clear. The stockholders own the company, and in their interest, the firm is required to maximize wealth. Given this responsibility, the appropriate role is to develop the investment opportunity in which a profit can be made. Since an energy company's expertise is basically in the resources business, that's where the majority of investments are likely to take place but, other opportunities will be considered. And, each investment made must carry with it the opportunity to make a profit which is commensurate with the risks involved. Many opportunities in the energy field today hang in a delicate balance where profit potential is present, but the threat of more arbitrary taxes or other government incursions is also present. In this regard, October 1981 marks a real milestone in the way energy market adjustments will take place in this country. At that time, the Emergency Petroleum Allocation Act, which enables current price and allocation controls will expire. If no new energy legislation takes its place and the DOE packs up and leaves, the outlook will indeed be brighter. Markets will adjust, and more economically efficient energy investments will be made.

15 Energy Independence— How? A Synthesis

Nake M. Kamrany and
Aurelius Morgner

In this chapter, an attempt is being made to draw the diagnostic analysis of the preceding chapters and to synthesize the various prognoses into a comprehensive framework for serious consideration. The various recommendations are organized as follows:

I. As a first step, we must be willing to confront the energy problem and to accept the fact that energy in the future will be more expensive than in the past. It follows that there is a need for consumers, producers, and policy makers to break with the old ways and adopt new directions. Clearly, this calls for new vision, intellectual and psychological reorientation, and re-examination of the old concepts regarding energy because results from the recent past have been nothing but calamitous, whether they are measured in terms of domestic ills of inflation and unemployment, or international weakening of the dollar and the enormous import bills. Energy had a major role in these problems.

The ill effects of our energy policy have not stopped at the economic dimensions alone. We have suffered equally in international prestige, global influence, and national psychological setbacks. Furthermore, we strongly believe that if we do not come to grips with the energy issue, it will lead us into further national frustration and possible military involvement abroad. However, the existing ills and the possibility of future calamities could be avoided if we take certain steps now that are feasible and would lead to energy independence.

II. As a second step, we must re-examine certain basic assumptions concerning energy which have been the basis of our energy policy over the last decade.

Until the end of 1980, the predominate view of the national leadership concerning energy independence was that of pessimism, i.e., energy independence cannot be achieved by the end of this century for the United States.

In November of 1980, the Senate Energy Committee released a report, "The Geopolitics of Oil", and took the position that it is impossible for the United States to achieve energy independence in this century. Senator Henry Jackson, Chairman of the Senate Energy Committee, stated that even if we do everything we can, we will still be importing oil at the end of this century. He further added that because of the continued heavy dependence on imported oil, linkages between energy, the economy, and national security will govern national policy in the 1980's.

Moreover, the records of the past administrations reveal that there was a lack of focus concerning energy policy. This explains the fact that our past policies have tried to do a "little of everything" without precipitating a major impact in any one direction to solve the energy problem. The underlying reasons for these misdirected approaches were predicated upon the following precepts which had formed the cornerstone of the energy policy.

1. It has been assumed that the rate of energy consumption must grow at the same proportion as the growth rate of the GNP. A direct and proportional dependence of the GNP growth rate upon energy input was a foregone conclusion. The implication of this fixity of relations between energy and GNP was that we had little flexibility in reducing energy consumption through conservation or the price mechanism since it would contribute to lower economic growth rate and the concomittant problems of unemployment would become greater. James Sweeney of Stanford University has refuted this assumption by showing that it is possible for the economy to grow at a constant or reduced rates of energy input including oil input. Under certain assumptions, a reduction in oil consumption by 40% may reduce the GNP by only 1%. Moreover, recent data (1979/80) has shown that the GNP has increased while the rate of energy consumption has declined in the United States and a number of industrialized countries.

2. There has been what we may call a fallacy of the price elasticity of demand for energy, including oil. It has been assumed that responsiveness to relative price changes of energy was so low that energy consumption would not respond to price changes. In other words, the price elasticity of demand was assumed to be inelastic in such a way that energy consumption needs of the households and industry were inflexible, and would not decline in response to a price rise. The policy implication of this assumption was that the market mechanism of price incentives was largely ineffective as a vehicle for involuntary conservation, as well as for allocation, distribution, and production of energy. Robert Pindyck of M.I.T. and others have shown a much higher elasticity coefficient than before—from 0.2 to 0.9. More significantly, he has shown a

much higher elasticity coefficient for the long run. Several other empirical observations have been made concerning more efficient use of energy, such as auto milage per gallon of gasoline and/or energy input per unit of the GNP. During 1980, oil consumption declined by 8.4%, largely due to price increases.

3. The third assumption was that the domestic supply of energy, including oil production, cannot be increased appreciably. The policy implication of this assumption was a heavy reliance upon imports. Richard J. Stegemeier has effectively challenged this assumption and has pointed out the enormous potential that exists in the United States for increasing the supply of energy both in the short and the long run. He has pointed out that substantial federal lands could be released for oil exploration. Moreover, in response to higher prices, substantial exploration for new oil has already taken place in North America. The record for 1980 in terms of oil exploration was unprecedented.

4. It has been assumed that the market incentive mechanism was inappropriate for the creation and development of new technologies that would produce alternative sources of energy. The policy implication of this assumption was that the government must get engaged in huge programs of direct investment and undertake major research and development efforts. Lester Thurow has challenged this assumption and has proposed a specific market incentive mechanism for the creation of alternative sources of energy via the private sector within 10 years. The government does not have the knowledge nor the ability to do what the private sector can do in this field. As Lester Thurow has pointed out, some segments of the economy may become richer. But if that is what we have to pay to get energy independence, let us accept the fact and get on with the work.

5. Finally, it has been assumed that maintaining cheap prices of energy would serve the distributional goals of the nation. Any price increases were assumed to create a heavier burden upon the poor than upon the rich. Maintaining a policy of cheap energy did not actually serve the distributional goal since cheap energy benefitted the middle income and the rich who consumed more energy per capita than the poor. Besides, there are alternative and more efficient mechanisms to enhance the distributional goal than by maintaining cheap energy. This mechanism has created many other costs, such as unfavorable balance of trade, inflation, and security problems.

Moreover, other considerations related to alternative sources of energy, such as environment aesthetics, pollution, and other social considerations were used to practically retard the development of coal and nuclear energy. These considerations were based more upon the emotional predilection of the anit-growth culture of the 70's than upon a realistic assessment of the problem. Maintaining a cheap energy policy has amounted to subsidising foreign consumption of U.S. agricultural and industrial products since both are highly energy intensive, largely due to subsidies of the energy input cost.

In addition to the above well-intentioned but misguided assumptions, other considerations, such as elections, lobbying, interest groups, lack of counter-vailing force, and a process of myopic optimization, have compounded the energy problem. From the preceding observations, it follows that the United States' energy policy for the 80's should distill and modify these assumptions with a view to creating a framework for energy independence.

III. New Trends Concerning Energy

The confusion of our energy policy and the lack of will to take constructive steps is best explicated in Carolyn Kan Brancato's thesis that all we have done is a little of everything. She points to several trends, however:

1. Pre-embargo market disruptions and a continued OPEC threat have stimulated a desire to achieve self-sufficiency through increasing supplies and reducing demand.

2. A direct Federal presence in energy markets has gradually become more acceptable. Cooperation between the public and the private sector is now common, especially where the government is the risk taker.

3. It is now more acceptable to price energy at its replacement value instead of subsidizing its cost. This pricing policy has the dual purpose of reducing demand, to the extent that it is price elastic, and of increasing supply. President Ronald Reagan's speeded-up decontrol of gasoline prices will probably curtail demand and expand domestic supply of oil.

4. In terms of dollar values authorized to increase energy supply or to reduce energy demand, it appears that the supply side is ahead, especially considering the large synfuels subsidy.

5. Our energy policy so far appears to be based on a "try a little of everything" approach with incremental gains in increasing supply rather than in reducing demand. The dispute over this policy is whether adequate consideration has been given to formulating this policy with respect to efficiency and established priority as to which method—for example, solar or synfuels—gives the greatest return for the federal dollar expended, and how to use the federal expenditures to reach the goal.

IV. Approaches Toward a Solution

The foregoing analysis pointed to some of the reasons for our inability to tackle the energy problem effectively. By tinkering with market mechanisms we created a much bigger problem than if we had left the market alone. Nevertheless, the complexity of the problem, its supranational character, its dynamic nature, and the fact that the energy problem resulted from several fallacious assumptions and a set of sequential and interacting chains of events depicting myopic optimization are among the causes for a lack of a national consensus concerning the origin and the nature of the energy problem and what must be done to resolve it.

A. CONSENSUS

There is some consensus among the various proposals:

1. There is no problem of physical scarcity of energy. In other words, there will be ample energy to sustain our basic way of life.

2. There is a clear distinction between the energy problem in the short run and the energy problem in the long run. It follows that approaches to tackling the short- versus the long-run problems are different.

a) The short-run approach essentially calls for adjustments to a changing and unstable supply and strategic costs. The most significant aspect of the adjustments is involuntary conservation through the price mechanism and incentives. However, it must be noted that lifting regulations and adhering to the market mechanism by itself is not sufficient to create energy independence. President Ronald Reagan's price decontrol of January 1981 may create the necessary conditions. But the sufficient conditions call for limits to imports and the creation of incentives to produce new supplies of energy.

b) A common thread among the proposals is the imposition of import limits upon oil as a response to the enormous burden of the oil imports upon the balance of merchandise trade and as a means of achieving conservation and rationality, and generating funds for investment in energy technology for the long-term solutions.

c) A concomitant issue of the short-run problem is the issue of energy security. The proposed strategy is to change vulnerability and major disruption into a minor supply adjustment through international agreements, and creation of new supplies. Stockpiling needs a hard look.

3. The long-run approach (10-15 years) relies upon the creation of new sources of energy through technologies, inventions, and innovations. Several mechanisms for technological developments have been suggested, including direct subsidies to encourage research and development, guaranteed markets and prices for new sources of energy, and government direct involvement for creating new sources of energy.

4. For the creation of new sources of energy, the private entrepreneurs are relied upon to respond to incentives via the market system, i.e., guaranteed prices and markets.

5. Much flexibility is possible in the relationship between energy and economic growth, thus allowing for better chances of smoothness in the needed adjustment.

6. Likewise, much more flexibility is possible in the consumption of

energy by the household consumers and industry users through involuntary conservation and substitution.

7. Revitalization of the market system, including price reforms, appears to be a pivotal point in these deliberations.

8. Finally, increased rationality and efficiency in energy/economic decision-making in the short-run and taking appropriate measures now to create long-term solutions with an explicit recognition of the time element for technological development of alternative sources of energy. These ideas have led to a number of complementary proposals for the short and long run.

B. BREAKING THE VICIOUS CIRCLE OF THE ENERGY PROBLEM

Kamrany derives his solution by defining the interacting energy problem, i.e., the vicious circle of the energy problem in the short run whose roots lie in the aggregate demand for energy. He advocates involuntary conservation in the short run by decontrolling the energy market that should reduce aggregate demand and oil imports. The long run solution relies upon technological development through the creation of new sources of energy. Kamrany expresses optimism about the long-run solution in view of the history of technological change, inventions and innovations which have responded to market demand. In his view, taking the government out of the energy market altogether, except for the creation of generic technologies and market incentives, will provide the necessary and sufficient conditions toward resolving the energy problem in the short run and in the long run.

C. CONFRONTING THE PROBLEM AS A SOLUTION

Roger Noll asserts that our basic problem is our reluctance to face facts. Energy is more expensive and will continue to increase in price. And, we have become dependent on an unstable source of supply for a large part of our energy, namely, the Middle East. We have to create domestic sources of energy and decide between coal and nuclear power for electricity in the next decade or two. We must conserve energy and develop new energy sources, such as synthetic fuels, for the year 2000 via private enterprise. Our future is one in which energy is more expensive, but ample to sustain our basic way of life. The sooner we face these facts, the quicker the energy crisis will end.

D. GUARDING AGAINST OIL SUPPLY VULNERABILITY

Vulnerability to oil supply disruption arises because we import so much of the oil from the politically unstable Middle East. While this problem is unpredictable, a significant disruption will have very severe economic and political impacts on our allies, as well as on ourselves. Since production of alternative liquid fuels is at least ten years away, we need contingency planning to prepare us to deal with possible supply disruption. The United States would suffer a loss of some $200 billion, or seven percent of the GNP from such disruption. Similar estimates for Japan and Western Europe will be nine to ten percent of their GNP. And oil prices will rise to above $100 per barrel.

Walter Baer identifies the following short-term or standby responses:

Preparatory measures for emergency supplies
- National petroleum reserves
- Private stockpiling of oil and refined products
- Standby production and distribution capacity
- Plans for emergency fuel switching

Preparatory measures for emergency demand restraint
- Standby gasoline rationing
- Emergency taxes or tariffs

E. DEREGULATION OF ENERGY MARKET AND A HIGH TAX ON GASOLINE

Robert Pindyck has identified the direct recessionary impact of a price increase of oil and energy and its indirect inflationary impact that would cause a reduction in the GNP and employment. His prescription for the American energy policy is complete deregulation of energy markets to avert energy shortages and reduce our dependence on insecure supplies of imported oil. Moreover, it is essential that we further reduce our dependence on imported oil, even beyond what would result from price deregulation. The simplest, most effective, and least costly way to reduce this dependence still further is by imposing a tax on gasoline. A large gasoline tax—around a dollar a gallon—would still make gasoline in the United States cheaper than in most European countries, and would represent a clear commitment to reduced import dependence. The proceeds of the gasoline tax could be used to reduce payroll taxes. This would reduce the total cost of production, and offset both the recessionary and inflationary impacts of rising energy prices.

F. REDUCE ENERGY USE BY REDUCING OIL IMPORTS

James Sweeney has addressed the issue of the effect of energy sector changes upon income and wealth of the nation. Specifically, the impact on the economy of changing availability or changing costs of energy has been evaluated. Structural unemployment occurs whenever the price of any factor of production increases quickly, energy included. Purchasing power is reduced; the demand for goods and services consequently goes down. These multiplier effects lead to additional short-run unemployment.

However, the long-run issues of economic growth with reduced availability or higher cost of energy are significant since there is a common belief (in government, industry, and public) that the economy cannot grow without roughly proportional growth in the quantity of energy use. This is a belief in a lock-step relationship between economic growth and energy growth. Many of our national energy policies have assumed such a belief. After considerable analysis, Sweeney concludes that we must distinguish between increases in the cost of imports or energy and variation in the availability of energy. He concludes:

IF THE COSTS OF PRODUCING ENERGY OR IMPORTING ENERGY INCREASE, THERE WILL BE A SIGNIFICANT IMPACT ON THE ABILITY OF THE ECONOMY TO GROW.

For instance, in 1979, the world oil prices increased by $15 per barrel. This cost increase for the United States amounted to $45 billion or roughly 2% of the GNP. That is a large number. If the 1979 price of imported oil doubles, it will cost 4% of the GNP. If we assume that energy commodities constitute 6% of the GNP, then if per unit energy cost increases by a factor of five, it would reduce GNP by 10% to 25%, depending upon the elasticity of demand.

On the other hand, restrictions on the availability of energy, not associated with cost increases, will have a far smaller impact on economic growth. For instance, a tax on the importation of all oil or an excise tax on its use will not increase the cost to the economy of importing oil but will increase the price people pay. If we reduce our energy use by 40% and the elasticity of demand were 0.9, it would cost one percent of the GNP.

It follows that if the elasticity of demand for energy is large, we can gradually reduce our use of energy, while imposing a very small proportional impact on the economy. While reductions in energy availability will reduce GNP without destroying or badly wounding the economy, a reduction in oil imports could actually increase GNP.

How? As we reduce the importation of oil, the world oil price may be reduced from what it would be otherwise. Thus, reduction in the use of energy may reduce its per unit cost, through reduction in the world oil price. This cost reduction may more than compensate the economy for energy availability reductions.

THUS, THE NET EFFECT OF ENERGY CONSERVATION MAY INCREASE GNP, IF IT LEADS TO REDUCTION IN THE WORLD PRICE OF OIL.

Sweeney concludes that this phenomenon leads to the concept of an import premium, i.e., a fairly significant tariff on the importation of oil. Depending upon the different assumptions, numbers between $5.00 and $70.00 per barrel may be the GNP maximizing tariff on oil importation. The revenues from import tax could be used to reduce corporate and personal income taxes which may lead to capital formation which is badly needed if we are going to improve our productivity.

In summary, energy cost increases can significantly reduce economic growth. However, reductions in energy use can be motivated through tax policy or other conservation programs without severly inhibiting economic growth. This policy would reduce vulnerability to oil supply disruption, reduce the world oil price, and reduce corporate and personal taxes which would lead to capital formation.

G. EXPAND THE SUPPLY OF ENERGY

Richard J. Stegemeir points to substantial potential for the expansion of the supply of energy. He estimates that only 10% of the world's known fossil fuel resources have been produced. The United States has approximately 30% of the world's supply of energy, consisting of predominantly coal (68%), oil shale (14%), and nuclear (9%). Only 9% of the United States energy supply is in gas and oil. Stegemeir identifies the energy problem as one emanating from a lack of national will which is embedded in political and socio-economic rigidities rather than technological constraints or resource scarcities. Government intervention of the energy market has created the energy problem.

His short- and long-term energy options are summarized below:

SHORT-TERM POLICY RECOMMENDATIONS — 1981-2000
1. REOPEN FEDERAL LANDS FOR OIL AND GAS
2. DECONTROL ENERGY PRICES
3. REMOVE DISINCENTIVES FOR COAL, NUCLEAR, & OIL SHALE
4. CONSERVATION

LONG-TERM POLICY RECOMMENDATIONS — NEXT CENTURY
1. FUSION
2. SOLAR
3. SYNTHETIC FUELS FROM COAL, BIOMASS

H. IMPOSE LEGISLATIVE LIMIT ON OIL IMPORTS

Delmar Bunn proposes an oil import limit as the most central element around which policy can logically develop. He proposes a two-tier tax system: First, a tariff should be legislatively set high to encourage the development of alternative sources of energy through the market mechanism, and conservation. Secondly, a variable tariff should be imposed to adjust the balance of merchandise trade at a healthy and acceptable level.

It follows that the import limit should be in dollars, not barrels. Revenues generated by the import tax would be used only for a precisely defined energy program, assistance in building local transportation systems, efficient energy utilization programs, and support for the energy administration until the existence of this program is no longer needed, perhaps ten years. Bunn maintains that such a program will improve the balance of trade; reduce the price of oil, thereby reducing the rate of inflation; create new employment through capital investment for domestic energy; and strengthening the dollar abroad. He believes it is politically feasible.

I. GUARANTEE MARKETS AND PRICES FOR NEW ENERGY

As a strategy for achieving energy independence, Lester Thurow advocates the adoption of an incentive scheme that guarantees markets and prices for those sources of energy which have come about in response to the incentive scheme. He believes that the development of alternative sources of energy — coal, shale oil, nuclear, solar — via a guaranteed price and market is the best mechanism to assure long-term supplies of energy.

Initially, the guaranteed price should be set high enough, say $50 per barrel, to effecuate a positive response. New projects that are created in response to this initial incentive must enjoy the benefits of the incentive plan, both in terms of price and market guarantees. Subsequently in a couple of years, the response should be assessed in terms of estimates of future supplies. If they are going to be above the needed supply, then the incentive measures, i.e., guaranteed price and markets, should be lowered as compared to the initial incentives, or vice versa.

This sort of assessment will go on intermittently, and appropriate adjustments will be made. The new guaranteed price and markets may be higher or lower than the previous ones, depending upon the estimates of future supplies. This strategy, according to Thurow, will bring about energy independence in 10 to 15 years.

For the short run, Thurow suggests involuntary conservation, either by 1) an imposition of a high tax on oil and a rebate mechanism of the taxes by lowering sales and or property taxes or 2) instituting rationing (coupons), plus a white market for the coupons to be bought and sold freely.

Either of the above approaches—tax or rationing—have their pluses and minuses. Both would require sacrifices and these sacrifices would have to be shared and accepted by our society at large if we are going to resolve the energy problem and achieve energy independence.

J. THE NATIONAL ENERGY DIVIDEND AS A SOLUTION

Michael D. Intriligator proposes NED for the imposition of substantial federal taxes or surtaxes on all final energy uses, including gasoline, heating oil, electricity, and natural gas. On gasoline, for example, an additional federal tax of 80 cents to $1.00 a gallon would be imposed, and comparable taxes would be imposed or added to existing taxes on other final uses of energy. The funds collected under NED would all be placed in an Energy Trust Fund, which would be established for this purpose.

The Energy Trust Fund would be disbursed in two ways: Five percent would be allocated to research and development or new and improved energy systems and on the reduction of undesirable impacts of energy use, such as environmental pollution. The remaining 95 percent would be returned to the public; all adult individuals with adjusted gross incomes of less than $25,000 per year, receiving an equal share.

The research funds would be disbursed through the existing federal funding agencies for research and development, and the 95 percent allocated to individuals would be paid on an equal basis regardless of energy use, location, etc. The Internal Revenue Service would disburse these funds. Through NED, substantial increases in the cost of using energy would lead to a more efficient utilization of scarce energy resources, conservation of these resources, and reduction in environmental pollution. Low- and middle-income families would not be adversely affected since they would share in the proceeds of the new taxes.

In sum, NED would:
1. ensure an efficient utilization of scarce resources
2. finance promising energy research and development projects
3. augment incomes of low-income families
4. provide an income for all individuals

K. FINAL SUMMARY FOR POLICY CONSIDERATION

We believe the following steps will provide the necessary and sufficient conditions for the United States to achieve energy independence:

1. **Decontrol of Prices.** The Administration's decontrol of gasoline prices, which was announced by President Ronald Reagan on January 28, 1981, in our view, provides for a constructive and necessary step toward energy independence for the United States. This measure will improve the efficiency of gasoline use by all sectors of the economy, including household and industrial, thereby, reducing the demand pressure for gasoline in the short run. Moreover, incentives will be created for domestic exploration and production of energy which will increase the supply side of energy in the intermediate and the long run.

However, we believe that the decontrol measures should be supplemented by a number of other measures in order to create sufficient conditions toward energy independence for the United States. These policy measures are discussed briefly below:

2. **Limit Oil Imports.** We recommend a dollar limit on oil imports and believe that such a policy will have a significant positive impact. It will reduce the aggregate demand for oil imports which, in turn, will depress oil prices in the international market since the United States is a major purchaser of oil. Indeed, this measure will be more effective upon the international price of oil if the rest of the consuming countries cooperate in limiting their oil imports. Such a concerted effort will reduce the OPEC prices or at least will stabilize them. In addition, the domestic economy will benefit as follows: A. The burden upon the balance of merchandise trade because of oil imports will be reduced which will result in strenghening the U.S. dollar abroad. B. The OPEC price stability will minimize the inflationary pressures of imported oil prices which have been significant in the past. C. As the price of imported oil is kept constant or reduced, it will have a net positive effect upon the GNP—it will increase since imports are subtracted from the GNP in the national income account. The positive impact upon the GNP will have a favorable impact upon domestic employment. For instance, a $1 billion reduction in the cost of imports could create about 50,000 jobs domestically.

The above policies may create a two-tier price system for oil. One, an unregulated market-determined price, paid by the consumers. This price will tend to rise toward the replacement cost for oil. And, two, an international price for oil which will be determined through bargaining among the consumers and OPEC. We anticipate that the domestic price paid by the consumers for oil will be higher than the international price for oil. Such a situation will be the reverse of what has been going on for the last nine years under the import subsidy and entitlement programs.

3. **Market and Price Guarantees.** While deregulation and limits on imports will have short-term effects upon aggregate demand for oil, and may encourage domestic production, including the development of technologies for alternative sources of energy in the long term, we believe that there is a need for providing an additional direct incentive to create new supplies of energy, i.e., guaranteeing markets and prices for new energy. This plan calls for a program of guaranteeing initial markets at some specified prices high enough to create incentives and avert risk for the private sector's investment. Then, every two years, such a program would be reviewed and modified in the light of responses to the program as measured in terms of actual and prospective new supplies.

4. **Conservation.** We believe that there is a need for an imaginative and effective national campaign to conserve energy and ease the transition from a "cheap energy" life style to one in which energy will be more expensive. Effective government-industry-university-citizen group col-

laborative efforts will be needed. Also, effective and imaginative use of the mass media campaign, conference, seminars, publications, and TV educational programs must be created to close the existing educational and information gap about the United States energy problem and to create a national consensus.

5. **Stockpiling.** We also believe that the cost of stockpiling reserves is prohibitive. Moreover, stockpiling will contribute to oil import demand which will strengthen, rather than weaken, OPEC. International agreements and other contingencies should be sought to divert a supply disruption problem into a minor adjustment rather than a major shock to the economy.

Index

Tariff
  to restrain demand, 46-47
  changes in, 65-66, 167-168
Tax
  on gasoline, 75
  on profits, 103
  to restrain demand, 149-150
Thurow, Lester, 23-27, 190
Trans-Alaskan Pipeline, 148

Unemployment, 73-74, structural, 54
United States energy growth rates, 53-55
United States Synthetic Fuels
  Corporation (SFC), 108-109, 131
Utah, tar sand, 130

Value, of the dollar, 73

"Vicious Circle", 1, 1-15, 168
Vulnerability
  reduction on, 142
  supply disruption, 141
  reduction of, 166

Wind Energy Conversion program, 134
"Wheeling", defined, 149
Weatherization Assistance Program, 128
Windfall profits tax, 116, 108-110,
  129-130
World energy supply, 21-22
Wyoming, coal and oil shale, 130

Yemen, 112, 126
Yom Kippur War, 142

About the Contributors

Walter S. Baer, Ph.D., is program director of Energy Policy Program at the Rand Corporation in Santa Monica, California. Dr. Baer's research has centered around the effects of government regulation and technological change in energy and telecommunications.

From 1967–1969, he served in the Office of Science and Technology in the Executive Office of the President with responsibilities for federal communication and computer policies. He was selected in a national competition as a White House Fellow and served with Vice-President Hubert Humphry (1966–1967). He received the B.S. from California Institute of Technology and the Ph.D. in physics from the University of Wisconsin.

Carolyn Kay Brancato, Ph.D., received the Ph.D. from New York University and the A.B. from Barnard College, Columbia University. She is currently section head, Energy and Industry Economics Section, Congressional Research Service, Library of Congress. Dr. Brancato's previous employment includes New York City Council, New York University, N.Y. City Government, Lehman College, N.Y. State Legislative Institute, Office of the Attorney General, N.Y. State, Arthur D. Little, Inc., National Science Foundation (project on energy conservation), and Dominick and Dominick, Wall Street Brokerage House.

Dr. Brancato has published numerous papers on the various aspects of energy, and is a contributing editor of the Empire State Report. She is appointed by the Governor to the Board of Trustees, College of Environmental Science and Forestry, State University of New York. She has also served on the Lieutenant Governor's Resource Council concerning Full Employment and Economic Development.

Delmar Bunn, M.D., is the executive director of the Newport Foundation for Study of Major Economic Issues. He has served both in Europe and the United States as educator, writer, speaker, and social counselor. His most recent writings in these fields include "A Basic Energy Program: Discussion of the Pivotal Move Necessary in any Effective Energy Plan: 1979;" "America Decides," 1980.

Dr. Bunn received the M.D. from the University of California in Los Angeles in 1964. He is in private practice in Newport Beach and is on the staff of the Hoag Presbyterian Memorial Hospital in Newport Beach, serving on numerous committees and in various positions of leadership including five years on the Executive Committee of that hospital. Dr. Bunn has

181

also undertaken graduate studies in the field of history at the University of Zurich, Switzerland, Heidelberg and Frankfurt, Germany.

David C. Button, M.A., is Getty Oil Company's manager of economic planning and policy, corporate planning department, headquartered in Los Angeles. Mr. Button joined Getty in 1969 as an economics analyst in Los Angeles. In 1972 he was promoted to investment planning supervisor and two years later was named economics supervisor. He assumed his present position in 1979.

Mr. Button received the B.A. in economics from Long Beach State College in 1967 and the M.A. in economics from the University of California at Los Angeles (UCLA) in 1969.

L.J. Cherene, Jr., Ph.D., is a Transportation Planner with Southern California Association of Governments and a Research Associate with the Department of Economics at the University of Southern California. He received the B.A. at St. Mary's College of California in 1971, majoring in economics. He continued his studies at the University of Wisconsin, Madison, earning the M.S. in 1974 and completing his Ph.D. in 1976. He spent the following year as a Leverhulme Fellow at the University of Hull, England. His experience in the field of energy economics includes modeling economic systems having fixed, nonreplaceable resources and constructing planning algorithms for capacity expansion of electricity generation. He developed a general equilibrium model of North America for OPEC's World Energy Model. He is currently involved with the Los Angeles Region Transportation Study's simulation of the effects of alternative land uses upon transportation demand.

Michael D. Intriligator, Ph.D., received the Ph.D. in economics at the Massachusetts Institute of Technology in 1963 and joined the faculty of the Department of Economics at UCLA, where he is now a professor of economics. He won Distinguished Teaching Awards for both graduate and undergraduate teaching. The theory and applications of quantitative economics, including mathematical economic theory and econometrics and their applications to health economics, strategy and arms control, and industrial organization, are Dr. Intriligator's major research interests. Among his several books in these areas are *Mathematical Optimization and Economic Theory* (1971); *Econometric Models, Techniques, and Applications* (1978); *A Forecasting and Policy Simulation Model of the Health Care Sector: The HRRC Prototype Microeconometric Model,* with Donald E. Yett, Leonard Drabek, and Larry J. Kimbell (Lexington Books, 1978); and *Strategy in a Missile War* (1967). He is also editor of *Frontiers of Quantitative Economics* (1971), Vol. II (1974), Vol. III (1977); and coeditor (with

Kenneth J. Arrow) of the *Handbook of Mathematical Economics* (1978). He has also published numerous articles in economics.

Aurelius Morgner, Ph.D., is professor and former chairman, Department of Economics, and professor, School of International Relations, at the University of Southern California.

Professor Morgner is a specialist in international trade and monetary theory and economic growth of the less-developed countries. He has served as an economic advisor to Brazil, Ecuador, the Philippines, the Yemen Arab Republic, and a number of international agencies. He is the author of books in the areas of international trade, economic development, and price theory.

Roger G. Noll, Ph.D., professor of economics and chairman of the Division of Humanities and Social Sciences, received the B.S. from the California Institute of Technology in 1962 and the Ph.D. from Harvard University in 1967. A member of the institute faculty since 1965, Professor Noll has also been a senior economist at the President's Council of Economic Advisors and a senior Fellow at the Brookings Institution. Professor Noll is the author of three books and numerous articles on safety and environmental policies, public-utility regulation, the broadcasting industry, the economics of professional sports, the application of economics in political science, medical-care policy, and bureaucratic decision making. His current research is on the development of a practical method for implementing tradable emissions rights as a means of dealing with pollution and on the creation of market-like mechanisms for small-group decision making that achieve better results than majority-rule voting.

Eileen Alannah Orrison, M.A., is currently a candidate for the Ph.D. in economics at the University of Southern California where she received the M.A. in economics. She received the B.A. from Immaculate Heart College in 1977.

Robert S. Pindyck, Ph.D., received the Ph.D. in economics from M.I.T. in 1971, and is now a professor of applied economics in the Sloan School of Management at M.I.T. He is author of *Optimal Planning for Economic Stabilization* and *The Structure of World Energy Demand;* coauthor of *The Economics of the National Gas Shortage: 1960-1980, Price Controls and the Natural Gas Shortage,* and *Econometric Models and Economic Forecasts;* and editor of *Advances in the Economics of Energy and Resources.* He has also authored or coauthored a number of journal articles on economic policy and the economics of energy and natural resources, as well as articles on energy and economic policy that have appeared in *Foreign*

Policy, The Public Interest, The Wall Street Journal, and *The New York Times.* Professor Pindyck has been a consultant to the Federal Reserve Board, the Federal Energy Administration, and the World Bank, as well as a number of private companies.

Richard G. Stegemeier, M.S., received the B.S. in petroleum engineering from the University of Missouri and the M.S. in petroleum engineering from Texas A & M University. In 1951 Mr. Stegemeier joined Union Oil Company of California as a research engineer and has held several positions in that company, including vice-president in charge of research, president of the Union Science and Technology Division and he is currently senior vice-president of Union Oil.

Richard J. Stone, J.D., is the director of the Department of Energy's (DOE) Office of Intergovernmental Affairs. Mr. Stone also provides day-to-day management oversight for the DOE Regional Offices and serves as chairman of the Department of Energy Council for Intergovernmental Affairs.

From 1978 until assuming his present position, Mr. Stone served with the Department of Defense (DOD) as the deputy assistant general counsel for Intelligence, International and Investigative Programs in Washington, D.C. Before joining the DOD, Mr. Stone served briefly as a special deputy to the county assessor for the County of Los Angeles.

Mr. Stone received the B.A. in economics from the University of Chicago in 1967 and the J.D. in 1970 from the University of California at Los Angeles School of Law.

James L. Sweeney, Ph.D., is a professor of engineering-economic systems at Stanford University. He holds the B.S. from M.I.T. and the Ph.D. in engineering-economic systems from Stanford University. He served as director of the Office of Energy Systems Modeling and Forecasting at the Federal Energy Administration while *The National Energy Outlook* was being written and while the supporting energy policy models were being developed. His articles have appeared in *Econometrics, Journal of Economic Theory, Journal of Urban Economics, Resources and Energy, Review of Economic Studies, Management Science,* and other journals and books. He currently serves as director of the Energy Modeling Forum, a national activity, headquartered at Stanford University, aimed at improving the use and usefulness of energy models.

Lester C. Thurow, Ph.D., is professor of economics and management at M.I.T. He is one of the leading social scientists in the field of economic reform, welfare economics, and the quality of life. He has authored and contributed chapters to thirty books. He has been on the faculty or staff of the Harvard University, Council of Economic Advisors, and various governmental and private groups. He is currently associate editor of the *Quarterly Journal of Economics and Review of Economics and Statistics.*

About the Editor

Nake M. Kamrany, Ph.D., is with the Department of Economics at the University of Southern California, where he also serves as director of the Program in Law and Economics. Previously, he was director of the Program in Productivity of Technology. He has held faculty and senior staff positions with the Massachusetts Institute of Technology, Stanford Research Institute, the World Bank, System Development Corporation, Battelle Institute, and UCLA. He specializes in applied economic policy, including energy, technology, economic analysis of law, and corporate strategic planning. He has authored and edited numerous books, research reports, articles, and position papers. His recent books include: *Economic Issues of the 80's; The New Economics of the Less Developed Countries* (1978); *International Economic Reform* (1977); selected articles include: "Technology, Productivity, and the U.S. Economy," MIT; "U.S. Productivity and Foreign Trade," National Inquiry into Productivity in the Durable Goods Industry," National Science Foundation; and "The Three Vicious Circles of Underdevelopment"; and "The Sahel Sudan Case of West Africa."

Dr. Kamrany received research grants from the National Endowment for the Humanities, National Science Foundation, Agency for International Development, and Advanced Research Project Agency. He has been a consultant to many private companies, governmental agencies, and international bodies.

Date Due